Bacteria, Plasmids, and Phages

Bacteria, Plasmids, and Phages

An Introduction to Molecular Biology

E. C. C. Lin
Richard Goldstein
Michael Syvanen

Harvard University Press
Cambridge, Massachusetts, and London, England   1984

Copyright © 1984 by the President and Fellows of Harvard College
All rights reserved
Printed in the United States of America
10 9 8 7 6 5 4 3 2 1

This book is printed on acid-free paper, and its binding
materials have been chosen for strength and durability.

LIBRARY OF CONGRESS CATALOGING IN PUBLICATION DATA

Lin, E. C. C.
    Bacteria, plasmids, and phages.

    Bibliography: p.
    Includes index.
        1. Microbial genetics.   2. Molecular genetics.
I. Goldstein, Richard N.   II. Syvanen, Michael.
III. Title.   [DNLM: 1. Bacteria—Genetics.   2. Bacteriophages—
Genetics.   3. Plasmids.   4. Gene expression
regulation.   QW   51   L735b]
QH434.L56   1984   576'.139   83-22784
ISBN 0-674-58165-2 (alk. paper)
ISBN 0-674-58166-0 (pbk. : alk. paper)

Preface

THIS BOOK GREW out of a set of lectures on bacteria and their viruses designed to provide first-year students at Harvard Medical School with the background to pursue the study of human pathogens. Thus, throughout the work we point out connections with topics of clinical interest where appropriate and describe the general features of bacterial structure, physiology, and metabolism, as well as the behavior of viruses.

A major aim of this volume, however, is to give a compact and balanced account of the molecular genetics of bacteria and their associated plasmids and phages. *Escherichia coli*, phage λ, and the F plasmid are featured because these models have been most extensively studied until recently. We hope to give the student some insight into how knowledge of the molecular biology of microorganisms is gained and how it is being applied in the emerging technology of genetic engineering.

Though we assume a modicum of exposure to general biochemical pathways, enzymatic catalysis, and the basic properties of macromolecules, we believe that this volume can be useful as a text for upper-level college students who wish to be initiated into microbiology at the molecular level. A set of questions in each chapter reviews the material presented and challenges the reader to more detailed inquiries, often in the context of laboratory experiments. A complete set of answers is provided at the back of the book.

We wish to acknowledge the generosity of our colleagues D. R. Fraenkel and B. D. Davis, who permitted us to draw from their lecture notes, and J. R. Beckwith, who made a number of helpful comments. Daniel R. Kuritzkes and Alison De Long read all the chapters from the point of view of a medical and graduate student respectively and offered useful suggestions. Yu-Mei Chen and Eric Johnson assisted with the proofreading. We are grateful to Sarah Monosson for the many hours she devoted to the preparation of this work. Finally, we have had the good fortune of being able to rely on Susan Wallace as a discerning and patient editor. We alone are responsible for any inaccuracy or inadequacy in this work, despite the competent and abundant help we have received.

E. C. C. L.
R. G.
M. S.

Contents

Bacteria, Plasmids, and Phages

1

Bacteriology: An Overview

THE MOST widely distributed and metabolically diverse group of organisms known is bacteria. Some species live in close association with eukaryotic organisms: the relationship can be symbiotic, as in the nitrogen fixation process, or destructive, as in processes of infectious disease. Many bacteria are ecologically important, helping to maintain the geochemical cycle. Because of their relative simplicity and amenability to experimental manipulation, bacteria and their associated genetic elements have provided some of the most fertile and elegant systems used in basic research. In many cases principles established in research with bacteria have had applications of profound importance, as in the development of life-saving vaccines and antibiotics. Research in more complex organisms, too, has often depended on ideas and techniques acquired in bacterial research.

Medical Microbiology

Infectious diseases, including those of bacterial etiology, have presented a tremendous burden to humanity. Until this century they were not only the leading cause of death but also the cause of epidemics that disabled and terrorized communities, sometimes determining the fates of armies and of nations.

The spread of certain diseases from one person to another long ago suggested the existence of invisible, transmissible agents of infection. Microscopic organisms (microbes) were not discovered, however, until Anton van Leeuwenhoek (1632–1723) made one-lens micro-scopes with sufficient magnification. Among the previously invisible particles that he observed in many natural materials, some could be recognized as alive because their motility was not random. Never-theless, these observations in natural history, though subsequently expanded by the development of the compound microscope, did not lead to an experimental science of microbiology until nearly two centuries later, when a methodology was developed for culturing various microbes in ways that avoided contamination.

Louis Pasteur (1822–1895) played a seminal role in this develop-ment. He demonstrated that certain microbes are able to live without air, and that different kinds of microbes are associated with different kinds of fermentation (suggesting biochemical specificity also in infectious diseases). Working on this assumption, Pasteur went on to develop important vaccines, such as the one against rabies. Finally, in decisively refuting the theory of the continual spontaneous gener-ation of life (on an experimental time scale) by excluding contamina-tion from ubiquitous organisms floating in the air, he played a role in the nineteenth-century debate over evolution. Pasteur's famous re-mark "dans les champs de l'observation le hasard ne favorise que les esprits préparés" ("chance favors only the prepared mind") has become a universal admonition for experimentalists. His work was crucial in allowing later investigators to isolate a great variety of microbes and to study their biochemical and pathogenic activities.

Robert Koch (1843–1910) systematized medical microbiology by perfecting the use of solid media, on which individual cells give rise to separate colonies and hence to pure cultures. He also formalized the criteria known as Koch's postulates for distinguishing a patho-genic agent from the adventitious organisms also recovered from lesions. These technical and theoretical advances led immediately to great scientific progress and are still important to clinical and experi-mental microbiology. Between 1879 and 1889 bacteriologists in Ger-many discovered the tubercle bacillus, cholera vibrio, typhoid bacillus, diphtheria bacillus, pneumococcus, staphylococcus, strep-tococcus, meningococcus, gonococcus, and tetanus bacillus. Even today new bacterial pathogens are being discovered—for example, the Legionnaires' disease bacillus in 1977, and virulent strains in species previously considered harmless, such as enteropathogenic strains of *Escherichia coli*.

The small size of bacteria, near the limit of resolution of the microscope, initially impeded analysis of their internal structure and function. Progress was therefore limited to identifying species and describing their growth requirements, excretion products (including toxins), and interactions with their environment. Studies of viruses were made difficult by their requirement of an intracellular environment for multiplication. The basic cycle of viral multiplication was first deciphered in bacteria, while animal virology was long restricted to descriptive studies in the living host. Today several bacterial viruses (bacteriophages) are understood in minute detail; they not only provide valuable background for the understanding of animal viruses but also serve as model systems for genetic and biochemical research.

Despite the importance of microbiology in medicine, it should be noted that the identification of the major pathogens, however exciting, led to the development of antisera or vaccines for only a small number of those organisms. Indeed, even before the offending organisms were identified, the development of widespread sanitation early in the nineteenth century, separating sewage from the water supply, probably played a larger role than vaccines in reducing the frequency of serious infectious disease.

Sanitation, in conjunction with the introduction of antibiotics in the 1940s, had a dramatic impact on the pattern of disease prevalence, shifting it toward degenerative and neoplastic diseases and toward infections due to viruses or antibiotic-resistant bacteria.

Classification of Bacteria

The purpose of biological classification (taxonomy) is twofold: to identify species, the basic units of classification; and to arrange these in a hierarchical family tree. The arrangement has been based traditionally on degree of shared and unshared phenotypic characteristics. With higher organisms multiple kinds of evidence are used to conclude that a given hierarchy is a "natural" phylogenetic tree, reflecting lines of evolutionary descent. Moreover, with organisms that reproduce sexually, the species can generally be sharply delineated: it is a group of organisms that interbreed in nature. Thus, in the course of evolutionary divergence within a species (usually owing to geographic separation) the accumulation of mutational differences leads first to different races (subspecies), which remain interfertile, and then to reproductively incompatible species; only rarely are intergrades observed. The evolution of reproductive barriers between species was essential in producing the wide variety of higher organisms that are found in nature; otherwise interbreeding between variants would constantly tend to homogenize the whole population.

With bacteria, however, the groupings in the conventional branching tree serve primarily to provide a determinative key, by which one can conveniently identify the species of any new specimen; phylogenetic relations are inferred but are less certain, since various characteristics (morphology, biochemical activities, ecology, and so on) do not necessarily diverge in parallel. Because these organisms multiply vegetatively rather than sexually, the interbreeding test cannot define a species. Indeed, the ability of individual genes to mutate and occasionally to be transferred independently suggests the existence of a continuous range of variation, with no basis for defining species. However, in practice, comparison of bacteria with respect to a large number of characters shows that they form clusters of biotypes, each with many common features but also with a range of variation. This pattern implies that random combinations of bacterial genes do not have equal evolutionary survival value: there must be strong selection for a balanced and coherent genome, though we cannot yet identify its detailed features.

The classification of bacteria into species thus has a rationale but also a large element of arbitrariness; taxonomists will differ in the weight that they attach to different characters, and in their tendency either to "split" (favor a large number of small groupings) or to "lump" (favor a small number of large groupings). Nevertheless, with most new isolates there is no difficulty in agreeing on a species assignment, measured against a standard type culture. The differences between strains in a species can have clinical significance. As techniques have evolved to allow recognition of subtle molecular differences, it has become increasingly clear that these are often correlated with differences in pattern or degree of pathogenicity.

With the advent of molecular genetics, comparisons between the DNA of two organisms has provided a much more direct index of evolutionary relation than those between phenotypes. The simplest measurement is the percentage of guanine and cytosine in the DNA composition: genomes of bacteria vary much more widely than those of higher organisms, ranging from 30% to 75% (vertebrate genomes range from 36% to 43%), although within a closely related group, such as the Enterobacteriaceae, the range is narrow. A much more sensitive test is the degree to which the DNA of one organism will hybridize with that of another, after melting and then annealing under standard conditions. However, bacteria have diverged so much that DNA hybridization occurs only between organisms already known to be very close (whereas the whole range of vertebrate DNA will hybridize). Ribosomal RNA is much more conserved in evolution than are most genes, and hybridization of this RNA with homologous or heterologous DNA measures evolutionary distance over a broader range. So far, however, DNA comparison is a research tool and not a method used in diagnostic laboratories.

In comparing bacteria we should note that the components of the cytoplasm tend to be highly similar or even identical in their monomeric units (though usually not in sequence), whereas the macromolecules of the outer surface (wall, outer membrane, capsule) vary widely in their composition, including many novel sugars and modified amino acids. These external variations give rise to different serological types within a species and evidently have great evolutionary value for the pathogen, since a host who has developed protective antibodies against one type can subsequently be infected by another type in the same species.

Bacteria and the Geochemical Cycle

Pathogens of plants and animals constitute only a small fraction of recognized bacterial species, and an infinitesimal fraction of the total mass of microbes on earth. The vast majority of bacteria are beneficial or even necessary to the ecosystem, playing an essential role in the geochemical cycle by which CO_2 and N_2 are converted to organic matter and then reconverted to mineral form. The ability to fix CO_2 probably evolved first in bacteria, but the process has been largely taken over by algae in waters and higher plants on land. The chloroplasts in algae and plants are believed to be derived originally from photosynthetic bacteria that infected evolving eukaryotic cells and became established as endosymbionts. In contrast, the fixation of N_2 by reduction to ammonia and the subsequent conversion to nitrate (which will not evaporate from soil) remains an attribute only of bacteria.

In the other half of the cycle, which degrades all dead plant and animal matter, bacteria have evolved an astonishing variety of metabolic patterns and choices of nutrients. Although a given species can utilize only a limited range of substrates, the microbial kingdom taken as a whole can convert *all* natural carbon- and nitrogen-containing compounds to CO_2 and N_2. They cannot, however, attack many synthetic compounds, such as fluorocarbons and polystyrene.

The activities of bacteria in the soil have given rise to the applied field of agricultural microbiology, while the variety of fermentation patterns has led to the development of numerous cultivated strains for the production of foods and industrial chemicals. Bacteria provide numerous examples of specialized patterns of metabolism. For instance, *Propionibacter* ferments lactic acid to propionate and CO_2, which accounts for the holes in Swiss cheese. *Bdellovibrio* burrows beneath the wall of another bacterium and grows at the expense of its cytoplasmic content. Sulfur bacteria, acquiring energy by oxidizing H_2S or S, can produce up to 5% H_2SO_4 in the medium. Magnetotactic bacteria use intracellular magnetite (Fe_2O_3) crystals to orient their swimming along the earth's magnetic lines of force. Presumably by this device the cells move down into the bottom mud, where they find food.

The extreme ecological niches to which bacteria have become adapted in evolution is extraordinarily wide ranging. Many of the highly specialized species belong to the subkingdom Archaebacteria, regarded as relics of a primitive group of prokaryotes that were also ancestors of the subkingdom Eubacteria, to which most known species of bacteria belong. (Archaebacteria have cell membranes of unusual lipid composition and lack peptidoglycan walls, a hallmark of Eubacteria.) Representative of these "primitive" bacteria include the methanogens that reduce CO_2 with H_2 to give CH_4 and water; halophiles that thrive in the Dead Sea and require a salt concentration of over 10%; and thermophiles that thrive in hot springs at 90°C. There are also bacteria, the baryophiles, that live in the depths of the ocean and require a pressure of several thousand atmospheres. Even more astounding is the recent discovery that around sulfide chimneys on the deep ocean floor, some bacteria live at temperatures approaching 300°C. (At 265 times atmospheric pressure seawater remains liquid at temperatures of about 460°C.) At the opposite extreme there are the psychrophiles that can live at temperatures down to −5°C and cannot tolerate temperatures above 20°C.

Despite the great diversity among bacteria in their chemical abilities to exploit different compounds for growth and in their physical tolerance of specialized environments, the basic mechanisms by which genetic information is replicated and expressed and the basic pathways in central metabolism are highly similar and not fundamentally different from those of all other organisms. For instance, the glycolytic pathway in lactic acid bacteria is identical to that in mammalian muscle. Thus, the *diversity* of special biochemical adaptations and the *unity* of the core mechanisms of growth and reproduction constitute the two sides of evolution: divergence from a common ancestry.

Bacteria and Molecular Genetics

Bacteria were long considered essentially bags of enzymes, and their adaptive variation was vaguely thought to be an environmentally directed response of a plastic, primitive kind of genetic makeup. The existence of discrete, mutable genes was first recognized in the early 1940s, when one-step spontaneous mutations were demonstrated. However, classical genetics depended not only on recognizing heritable differences between individuals but also on carrying out crosses that yielded recombinants between two different parents. Because bacteria multiply vegetatively rather than by sexual reproduction, bacterial genetics progressed little until, in 1944, O. T. Avery's studies of pneumococcal transformation showed that genes can be transferred between bacteria and that the "transforming principle" is DNA. Soon thereafter the discovery of conjugation by Joshua Lederberg further showed that in bacteria, as in higher organisms, the genes are linked in a chromosome.

With the growing recognition of the fundamental unity of core metabolic pathways and hereditary mechanisms in bacteria and higher organisms it became possible to use bacteria as model cells for studying universal properties. Among the advantages of a bacterial system, the greatest is the ease with which one can cultivate billions of individuals and then select from these huge populations rare mutants and rare genetic recombinants. This is also true for molecular genetic studies in bacteriophage, in which the numbers of individuals per unit volume of culture can be even larger.

Avery's discovery provided an essential ingredient for developing a rich science of bacterial genetics; but even more, by identifying the genetic material as DNA, it laid the groundwork for the emergence of molecular genetics in 1953, with James Watson and Francis Crick's model of DNA structure. Intensive studies since then have revealed many aspects of DNA that contribute to its genetic and evolutionary functions: replication, expression, mutation, recombination, repair, and rearrangement. Moreover, viruses — and also plasmids (autonomous blocks of DNA in bacteria) — can integrate into the host cell chromosome and exchange genes with it. This mode of transfer provided powerful tools for studying and manipulating genes. The possibilities are now greatly expanded by molecular recombination in vitro, using enzymes from bacteria to insert DNA of any origin into viral or plasmid DNA, and amplifying the transplanted DNA (and often their products) by having the genetic vectors multiply in bacteria or higher cells.

Similar mechanisms for interspecific recombination are active in nature, where they play an important evolutionary role in microbes and almost certainly also in higher organisms (for example, the transfer of genes from a bacterial plasmid into plant cell chromosomes causes crown gall tumors). Indeed, this kind of gene transfer across species barriers provides more direct evidence than the fossil record for the evolutionary continuity of the living world.

In probing into the mysteries of the gene, molecular geneticists have not revealed any novel chemical or physical forces, as some people expected. However, the study of molecular genetics *has* revealed a pattern of organization that involves these forces in a way that is unique to living matter: **molecular information transfer.** Three kinds can now be recognized. First, DNA can store information, pass it on to future generations by replication of the sequence of bases in the DNA, and express the information by providing a template for complementary RNA sequences. Second, translation of the one-dimensional molecular information of DNA into the three-dimensional shapes required for phenotypic expression occurs simply by spontaneous folding and aggregation of RNA and polypeptide chains as dictated by the mutual affinity of their parts. Finally, allosteric proteins also carry out information transfer: they sense a chemical concentration and transfer this information to sites where it regulates the activity of a gene, an enzyme, or a protein involved in a sensory response.

Knowledge of genetic phenomena first revealed in bacteria has been extended to advance the understanding of human diseases. For example:

Single-gene defects   There are over 2,000 known human genetic disorders. Growing out of studies of biochemical defects in microbes, research on humans has now defined several hundred genetic disorders in terms of an altered or absent protein. Many of these disorders can be diagnosed prenatally by amniocentesis, followed by culturing the recovered fetal cells and identifying a defective protein. Moreover, DNA sequencing has recently been used to reveal alterations in genes coding for hemoglobin, for though the cells from amniotic fluid do not make the product, they have the gene.

Regulatory diseases   Gout is analogous to a class of bacterial mutant phenotypes caused by a genetically determined failure of feedback regulation of an enzyme, leading to overproduction of purines. More subtle variations in regulation of the amount or the activity of a specific enzyme or other protein (such as cell receptors) are being found in many other diseases.

Cell differentiation   The rearrangement of DNA segments during the maturation of lymphoid cells is responsible for the generation of innumerable kinds of antibodies. Mobile genetic elements in bacteria, which are inserted or recombine with high frequency at certain DNA sequences by special site-specific enzymatic mechanisms, provide a model, and such rearrangements seem likely to occur more broadly in cell differentiation.

Further Reading

Blakemore, R. P., and R. B. Frankel. 1981. Magnetic navigation in bacteria. *Scientific American*, December: 58–65.

Brock, T. D. 1978. *Thermophilic Microorganisms and Life at High Temperatures*. New York: Springer-Verlag.

Buchanan, R. E., and N. E. Gibbons. 1974. *Bergey's Manual of Determinative Bacteriology*. 8th ed. Baltimore: Williams and Wilkins.

Burnet, M., and D. White, 1978. *Natural History of Infectious Disease*. Cambridge: Cambridge University Press.

Dobell, C. 1960. *Antony van Leeuwenhoek and His "Little Animals."* New York: Dover.

Dubos, R. J. 1950. *Louis Pasteur: Free Lance of Science.* Boston: Little, Brown.

Friedman, S. M., ed. 1978. *Biochemistry of Thermophily.* New York: Academic.

Zinsser, H. 1963. *Rats, Lice and History.* Boston: Little, Brown.

2

Bacterial Structure

THOUGH ALL bacteria are small, different species have different characteristic sizes. For example, *Bacillus anthracis*, the cause of anthrax, is considered large (rods of about 1×5 microns; 1 micron $[\mu] = 10^{-3}$ mm), whereas *Hemophilus influenzae*, the cause of meningitis in children, is thought of as small (rods of about $0.3 \times 1\mu$). The smallest bacteria are probably the minimum size that allows the degree of organization necessary for independent metabolism. By contrast, subcellular infectious agents, such as viruses, which depend on their host cells for metabolism and reproduction, range from the size of small bacteria (as in the virus that causes smallpox) to particles 100 times smaller (such as the polio virus; see Fig. 2.1).

Even within the same species, bacteria show considerable variation in size. Thus, a newly divided cell is half the size of a cell just about to divide. Furthermore, rod-shaped cells such as *Escherichia coli* growing rapidly in a rich medium are much larger than cells of the same species growing slowly in a poor medium (Table 2.1).

In addition to a range in size, individual bacteria come in a variety of shapes: spheres (also called cocci), rods (bacilli), and curved rods (vibrios and spirilla). Most bacteria are rod-shaped, though curved rods are relatively rare. Sometimes bacterial cells are found in aggregates — cocci in pairs (diplococci), chains (streptococci), bunches (staphylococci), or cubical packets (sarcinae). These aggregates are the consequence of the cell division and the lags between separation of daughter cells, and aggregate shapes reflect the axes of successive cell division (Fig. 2.2). When viewed through the light microscope,

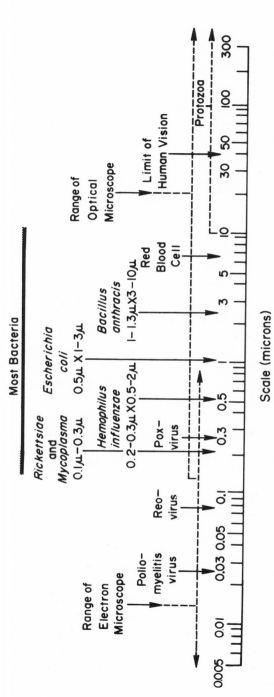

Fig. 2.1 Spectrum of microbial sizes. (From W. K. Joklik and D. T. Smith, eds. 1972. Zinsser Microbiology. 15th ed. New York: Appleton-Century-Crofts, p. 27.)

aggregates are often helpful in identification of the organism. In fact, some genera of bacteria were originally named according to the shape of their aggregate (*Staphylococcus*), others according to the shape of the individual cell (*Vibrio, Bacillus*). But the names of most bacteria say nothing about their shape.

The Cytoplasm

The bacterial cytoplasm consists of an aqueous solution containing hundreds (or thousands) of proteins of varying complexity, together with hundreds of metabolites. This intracellular material surrounds the DNA (deoxyribonucleic acid) and the main internal organelles, the ribosomes, where protein synthesis takes place (Fig. 2.3). In a number of species there are also cytoplasmic granules or inclusion bodies (occasionally containing storage molecules such as glycogen) whose staining properties are sometimes used to identify an infecting organism in the diagnosis of disease. Other membrane-bound organelles of eukaryotes are absent in bacteria: there are no mitochondria (widely believed to be themselves endosymbiotic bacteria in the remote past), endoplasmic reticulum, Golgi apparatus, or lysosomes. Many of the functions performed by these organelles in eukaryotes are executed by the membrane or envelope that surrounds the cytoplasm in bacteria.

Table 2.1. Composition of *Salmonella* at various growth rates.

Medium	Doubling time (min)	Dry wt/10^9 cells (μg)	Chromosomes/cell	RNA/10^9 nuclei (μg)
Minimal, lysine as sole carbon source	96	240	1.1	22
Minimal, glucose as sole carbon source	50	360	1.5	31
Broth	25	840	2.4	6.4

Source: O. Maaløe and N. O. Kjeldgaard. 1966. *Control of Macromolecular Synthesis: A Study of DNA, RNA and Protein Synthesis in Bacteria*. New York: W. A. Benjamin.

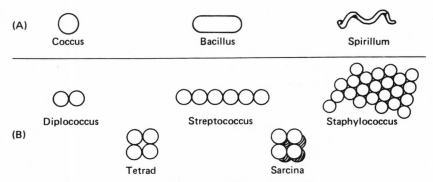

Fig. 2.2 Bacterial morphology. (A) Various shapes of bacteria. (B) Various arrangements of cocci. (Adapted from R. F. Boyd and B. G. Hoerl. 1977. *Basic Medical Microbiology*. Boston: Little, Brown, p. 13.)

THE DNA

The bacterial genome appears to be primarily a large single piece (**chromosome**) of DNA which is circular. There is no chromatin-like structure analogous to that of eukaryotes, nor is there a nuclear membrane; hence bacteria are called prokaryotes. Chromosomal DNA constitutes about 2% of the weight of a bacterial cell. A typical bacterial chromosome contains about 5×10^6 base pairs (the average molecular weight of one base pair is 660); the composition and sequence of base pairs varies with the species. A DNA chain of 10^6 base pairs has a length of more than 1 mm, or more than 1,000 times the length of the bacterium itself, so this DNA must be very tightly folded in the cell.

The chromosome is located in an area of the cytoplasm known as the **nucleoid,** distinguishable in the light microscope with certain stains or by electron microscopy in thin section. The chromosome is thought to be attached at some point to the cytoplasmic membrane or to its invagination, called a **mesosome,** whose function is uncertain.

RIBOSOMES

These are the nucleoprotein particles (composed of proteins and ribonucleic acids, or RNAs) on which protein synthesis occurs. Their complex structure is homologous but not identical to that of ribosomes in eukaryotes. The difference is very important, for some

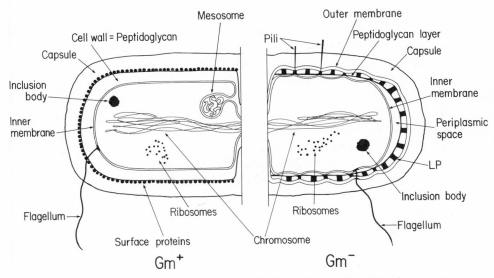

Fig. 2.3 Schematic representation of idealized bacterial cell and its appendages. The cell wall of gram-positive bacteria is shown on the left, gram-negative on the right. (From W. K. Joklik and D. T. Smith, eds. 1980. *Zinsser Microbiology.* 17th ed. New York: Appleton-Century-Crofts, p. 28.)

antibiotics specifically block bacterial protein synthesis by interfering with ribosomal function but do not affect protein synthesis in higher cells, including those of humans. Ribosomes are mostly associated with long chains of RNA, which are transcripts of the cell's genetic blueprint as written in the DNA.

The Cytoplasmic Membrane

The membrane which immediately encases the bacterial cytoplasm is known as the cytoplasmic or inner membrane. It is very similar both in appearance and in certain functions to other biological membranes. The bacterial cytoplasmic membrane is made up primarily of phospholipids and proteins, as is true for animal cell membranes, but it differs from them in having a higher protein content (60–70%) and in usually not containing sterols. In many bacterial species the major phospholipid is phosphatidylethanolamine. The degree of saturation of the phospholipids is finely adjusted to fit the environmental temperature; at lower temperatures the phospholipids contain more unsaturated fatty acids. Steric distortions introduced by double bonds help the lipid membrane to maintain its fluidity at low temperatures, so that a significant part of the membrane always exists in a "melted" state (that is, a two-dimensional solution).

The functions performed by the cytoplasmic membrane are many. First, it is the principal permeability barrier of the cell. It is essentially impermeable to all charged substances, even H^+. Only hydrophobic molecules or uncharged molecules no larger than glycerol can diffuse through this lipid bilayer. Thus, essential metabolites are not readily lost from the cytoplasm. The cytoplasmic membrane contains proteins that play roles in energy production, transport of metabolites into the cytoplasm, and biosynthesis of lipids and other cell envelope components.

The Peptidoglycan Layer

Completely surrounding the cytoplasmic membrane is a rigid net known as the peptidoglycan layer. This complex molecule, responsible for cellular shape and strength, is similar in nearly all bacterial species in both structure and biosynthesis. Because it is made up of a number of building blocks that are unique to bacteria, this polymer (also known as mucopeptide, or murein) is a good target for chemotherapy.

The essential structural features of peptidoglycan are a backbone of alternating residues of N-acetylglucosamine (GlcNAc) and its lactyl ether, N-acetyl-muramic acid (MurNAc), connected by β-1, 4-glycosidic linkage as shown in Figures 2.4 and 2.5. The GlcNAc residues are unsubstituted. The MurNAc residues are connected to a tetrapeptide of alternating D- and L-amino acid residues attached to the COOH group in amide linkage. Parallel strands of the polymers are cross-linked, usually via a peptide bridge. In the example shown in Figure 2.6, the peptide bridge is a pentaglycine, and it connects the terminal D-alanine residue of one strand to the ϵ-NH_2 group of the penultimate lysine residue of the parallel strand.

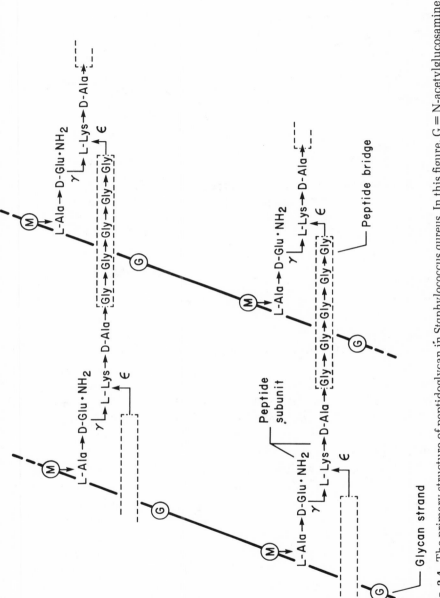

Fig. 2.4 The primary structure of peptidoglycan in *Staphylococcus aureus*. In this figure, G = N-acetylglucosamine and M = N-acetylmuramic acid. Arrows indicate the CO to NH direction of the linkages. Usual α-peptide bonds are represented by horizontal arrows; other peptide bonds (for example, γ or ϵ) are also indicated. The pentaglycine bridges, which extend from the ϵ-amino group of L-lysine on one peptide subunit to the carboxyl group of D-alanine on another, are enclosed by a dashed rectangle. (From J. M. Ghuysen and D. Stockman. 1973. In *Bacterial Membranes and Walls*, ed. L. Leive. New York: Marcel Dekker, p. 43.)

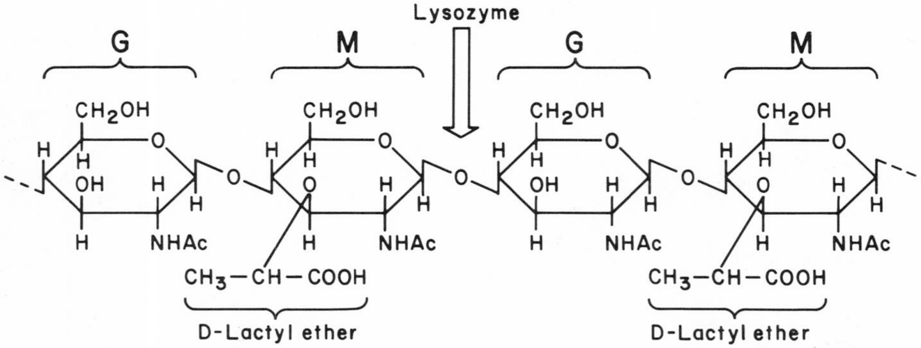

Fig. 2.5 A portion of a glycan strand. In the peptidoglycan network, the COOH of the D-lactyl groups is usually peptide substituted. The arrow indicates the site of lysozyme action (From J. M. Ghuysen and D. Stockman. 1973. In *Bacterial Membranes and Walls*, ed. L. Leive. New York: Marcel Dekker, p. 43.)

BIOSYNTHESIS OF PEPTIDOGLYCAN

While the biosynthesis of peptidoglycan is complicated, it is useful to acquire some feeling for how bacteria accomplish this unique construction. First, note that in Figure 2.6 the reactions are ordered not only in sequence but also in space, taking place in three different cellular locations. The basic building blocks of the structure, GlcNAc- and MurNAc-L-ala-D-glu-L-lys-D-ala-D-ala, are synthesized in the cytoplasm in UDP-activated forms. The amino acids are added by enzymes. This is an example of peptide-bond formation that does not occur on ribosomes. These building blocks are then transported through the cytoplasmic membrane by a polyisoprenoid carrier. Once externalized, these units are attached to the preexisting peptidoglycan template. A final interesting point is the formation of the cross-link. This peptide bond is formed, without any additional input of energy, at the expense of the breaking of an existing D-ala-D-ala peptide bond and the release of one D-ala residue (Fig. 2.7). In S. aureus the D-ala to D-ala peptide bond (CO—NH) is thus replaced by a similar D-ala to gly bond in the transpeptidation, or cross-linking, reaction. Other species use other cross-links.

Antibiotics that inhibit peptidoglycan biosynthesis are particularly useful, since animal cells do not have this protective wall and therefore are not affected. The classic example is penicillin. If this antibiotic is added to a growing culture of E. coli, lysis (dissolution) of

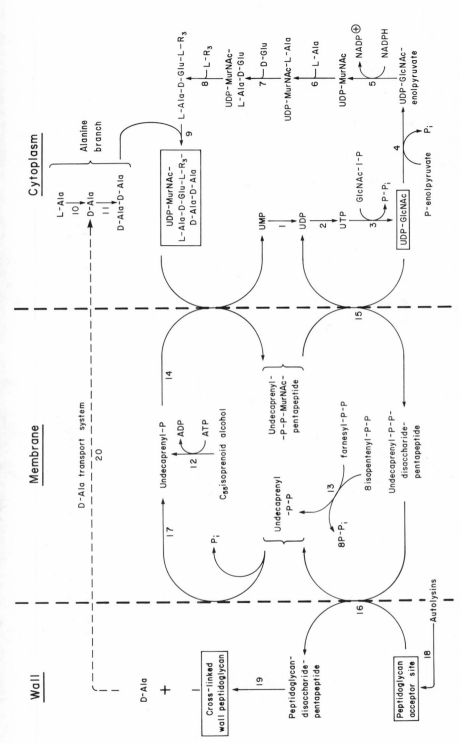

Fig. 2.6 Schematic representation of the biosynthesis of a peptidoglycan of chemotype 1. The three stages—cytoplasmic, membrane-bound, and wall-bound—are separated by the dashed vertical lines. All the reactions are shown in the anabolic sense. GlcNAc = N-acetylglucosamine; MurNAc = N-acetylmuramic acid. Residue L-R₃ may be: L-alanine, L-homoserine, L-diaminobutyrate, L-glutamate, L-lysine, L-ornithine, LL-diaminopimelate, or meso-diaminopimelate. (From J. M. Ghuysen and G. D. Stockman. 1973. In *Bacterial Membranes and Walls*, ed. L. Leive. New York: Marcel Dekker, p. 43.)

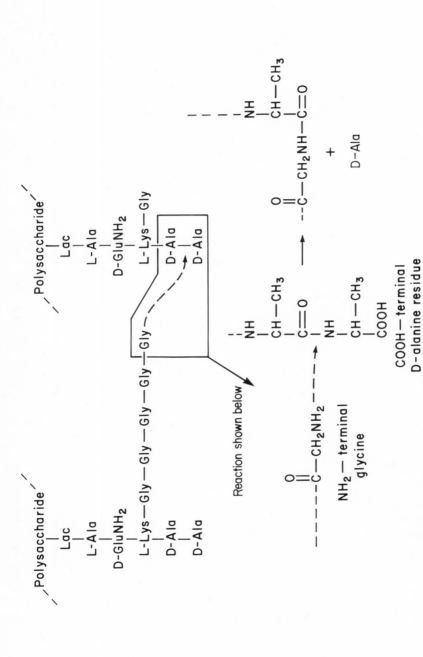

Fig. 2.7 The penicillin-sensitive transpeptidation reaction in *S. aureus*, which completes the cross-link between different peptide side chains: the D-ala to D-ala peptide bond (CO—NH) is replaced by a similar D-ala to gly bond. Some species use cross-links other than the pentaglycine bridge. (From B. D. Davis, R. Dulbecco, H. N. Eisen, and H. S. Ginsberg. 1980. *Microbiology.* 3rd ed. New York: Harper and Row, p. 80.)

the bacterial cells can be readily observed. If the same is done in media of high osmotic strength, lysis is prevented and protoplasts — bacteria without walls — are formed. The elucidation of peptidoglycan biosynthesis was accomplished during the course of searching for sites of action of penicillin, and it was eventually shown that one of the targets was the cross-linking step, transpeptidation.

All antibiotics that inhibit peptidoglycan biosynthesis cause lysis of the bacteria. Other examples are cycloserine, bacitracin, and phosphonomycin or fosfomycin (see Chapter 13). However, the connection between the inhibition of peptidoglycan biosynthesis and cell lysis is complex. Thus, in *Streptococcus pneumoniae*, the failure to cross-link peptidoglycan permits the escape from the cell of a teichoic acid that is an inhibitor of a peptidoglycan hydrolase, and this enzyme becomes active as an autolysin. Mutants lacking this enzyme are not killed by penicillin, though they cease growing.

If bacteria possessed solely peptidoglycan, they would be protected against osmotic lysis. However, murein hydrolases like lysozyme are ubiquitous, and these enzymes would quickly hydrolyze the unprotected peptidoglycan and cause cell lysis. This is probably one of the reasons why most bacteria possess additional external fortifications.

The Cell Wall in Gram-Positive and Gram-Negative Bacteria

Although the term "cell wall" is often used to refer to the peptidoglycan layer alone, strictly speaking the cell wall has many layers, only one of which is composed of peptidoglycan. (The cytoplasmic membrane and cell wall taken together are known as the **cell envelope**.) Studies of the cell wall have resulted in a division of the bacterial world into two major classes: the gram-positive bacteria and the gram-negative bacteria, so called because of differences in the response of these two types to Gram's stain. Gram's stain was devised to render bacteria more visible in light microscopy and has since become the most commonly used bacterial staining system.

In Gram's stain, a bacterial suspension is allowed to dry on a microscope slide and fixed by brief heating in a flame. The film is exposed to a basic dye (such as crystal violet) for a minute, and then to iodine, a mordant which fixes the dye to stainable material. Unfixed dye is removed by treatment with alcohol. The last step is treatment with a red basic counterstain, such as safranin. The preparation is blotted and examined without coverslip by light microscopy using the oil immersion lens.

Cells still blue after the decolorizing step are called gram-positive; the red counterstain does not change this color. Cells destained by alcohol treatment acquire the color of the red counterstain: they are gram-negative. This dissimilarity in staining properties is correlated with fundamental differences in the structure of the cell wall.

Classification as gram-positive or negative does not fit all bacteria. Some do not stain and have an unusual envelope. One such family is the Mycobacteriaceae, which includes the agents of tuberculosis. Mycobacteria have a very high content of unusual lipids, including waxes containing huge (C80) fatty acids, in their walls. They are

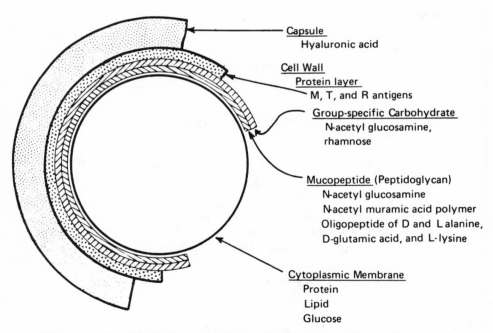

Fig. 2.8 A simplified schematic diagram of the cell envelope of a group A hemolytic *Streptococcus*. Although this diagram may not be correct in its detail, it does demonstrate how complex gram-positive envelopes can be. (From B. D. Davis, R. Dulbecco, H. N. Eisen, and H. S. Ginsberg. 1980. *Microbiology*. 3rd ed. New York: Harper and Row, p. 615.)

stained by an "acid-fast" method, such as heating over a steam bath with the dye carbol-fuchsin. Certain Corynebacteria, which include the agent causing diphtheria, are also acid-fast. Spirochetes (such as *Treponema pallidum*, which causes syphilis), thin spiral bacteria with unique gross structural properties different from other bacteria, also do not stain by the Gram technique.

GRAM-POSITIVE CELL WALLS

It is known that the peptidoglycan layer is typically much thicker in gram-positive than in gram-negative cells, but the structure of the outermost layer in gram-positive bacteria has not yet been elucidated in detail (Fig. 2.8). We do know that a number of clinically important antigenic determinants are located in this portion of the gram-positive cell.

Teichoic acids These compounds are a class of surface antigens found only in gram-positive bacteria. They usually have a poly-glycerol-P or polyribitol-P backbone substituted with a variety of molecules (sugars, amino-sugars, alanine; see Fig. 2.9). There are two general classes of teichoic acids. In one class, the teichoic acids are major antigens of the cell and are covalently attached to the peptidoglycan. Staphylococcal polysaccharide A is such a teichoic acid.

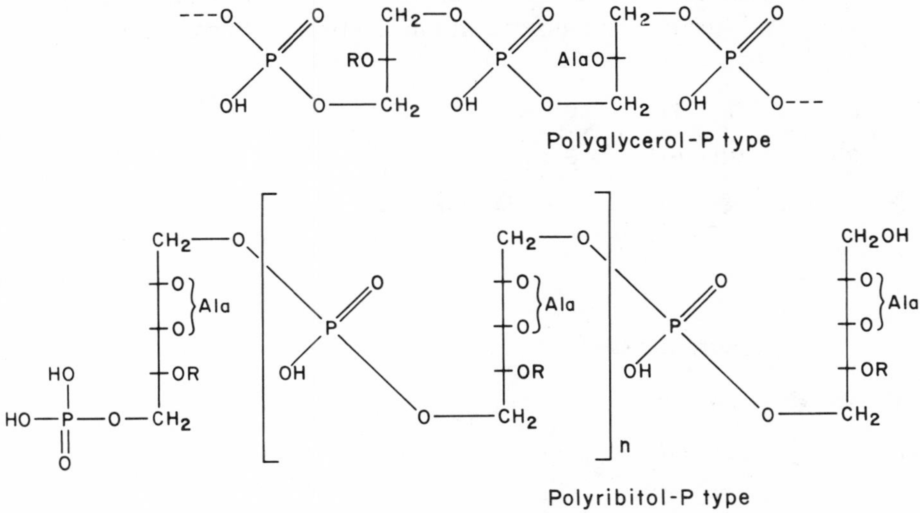

Fig. 2.9 Structures of teichoic acids.

The other class, lipoteichoic acids, is associated with the cytoplasmic membrane of all gram-positive bacteria. These compounds always consist of polyglycerol-P with a glycolipid at one end, typically a glyceryl disaccharide with two fatty acids on the glycerol (Fig. 2.10). In a few instances, lipoteichoic acid is also a major antigen (for example, *Streptococcus faecalis* group D carbohydrate).

Other polysaccharides in the gram-positive envelope can also be important antigens. For example, certain group-specific C (carbohydrate) antigens of streptococci are rhamnose-containing polymers. Streptococci also have a protein antigen (M protein) present in the envelope as pili (see below).

GRAM-NEGATIVE CELL WALLS

Gram-negative bacteria have a more complex cell-wall structure than gram-positive bacteria (Fig. 2.11). Between the cytoplasmic (inner) membrane and the peptidoglycan layer there is a compartment known as the **periplasmic space,** which is not found in gram-positive cells. Furthermore, in gram-negative bacteria there is a second membrane external to the peptidoglycan layer; it is referred to as the outer membrane.

The outer membrane   Although this membrane looks like the cytoplasmic or inner membrane in the electron microscope, the two

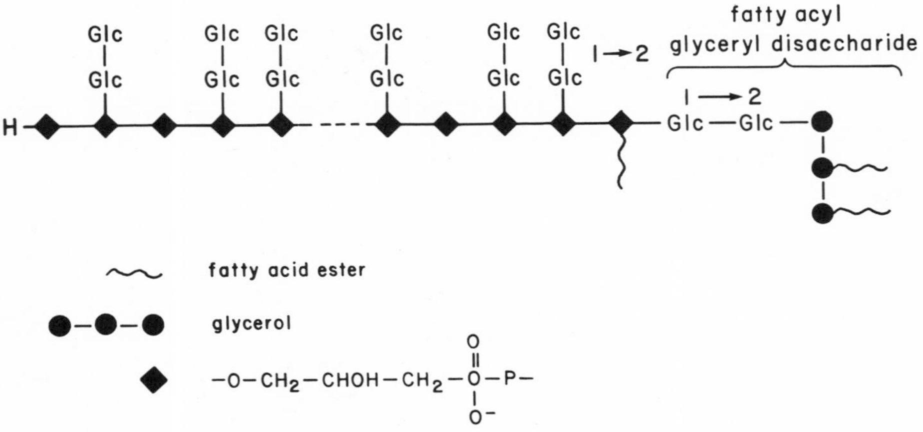

Fig. 2.10 Schematic representation of a lipoteichoic acid.

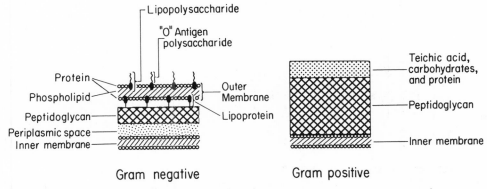

Gram negative   Gram positive

Fig. 2.11 Schematic representation of gram-negative and gram-positive bacterial cell envelopes. All layers external to the inner (cytoplasmic) membrane are often collectively referred to as the cell wall.

are markedly different in composition and function. The outer membrane contains both proteins and phospholipids like the inner membrane, but unlike it the outer membrane contains only a few kinds of major proteins. The outer membrane also contains a unique component, lipopolysaccharide (LPS).

The LPS molecule can be divided into three main parts (Fig. 2.12). Lipid A, embedded in the outer membrane, is a phospholipid based on glucosamine rather than glycerol. This lipid is the toxic component of endotoxin, the cell-associated toxic component of gram-negative bacteria. The next portion, the core polysaccharide, is a fairly constant structure in different gram-negative bacteria. Outermost, attached to the core, are specific polysaccharide side chains which are the major antigenic determinants in many gram-negative bacteria; these are called the O antigen. As an example, the O side chain of *Salmonella typhi* is shown in Figure 2.13.

The biosynthesis of lipopolysaccharide is analogous to that of peptidoglycan: the basic units of the molecule are made internally, transferred across the membrane barriers with the aid of the same polyisoprenoid carrier, and then finally assembled. The core is made by stepwise addition of activated sugar units to lipid A, whereas the O antigen is formed by polymerization of a tri- or tetrasaccharide intermediate. The mechanism of transport and assembly in the outer membrane is not yet understood.

The phospholipid bilayer of the outer membrane is asymmetric: LPS is located entirely on the outer leaflet; the inner leaflet is richer

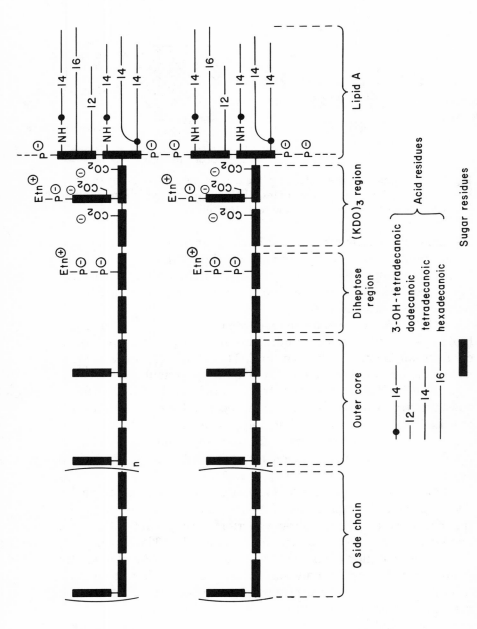

Fig. 2.12 Schematic representation of the lipopolysaccharide structure. (Adapted from H. Nikaido. 1973. In *Bacterial Membranes and Walls*, ed. L. Leive. New York: Marcel Dekker, p. 146.)

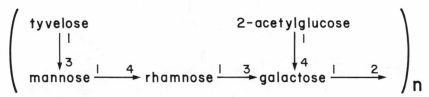

Fig. 2.13 O side chain of *Salmonella typhi.*

in ordinary phospholipids. One of the major outer membrane proteins, often called the **matrix protein** (porin), is transmembranal and appears to create a pore through the membrane (Fig. 2.14). This pore allows diffusion across the outer membrane of hydrophilic molecules whose molecular weights are less than 600. Another major outer membrane protein, the lipoprotein, contains on its amino terminus a lipid which probably helps to embed the protein in the phospholipid bilayer. Most of the lipoprotein molecules are covalently attached, via amide linkage with the carboxyl-terminal lysine, to the peptidoglycan, thus anchoring the outer membrane to the rest of the cell wall.

One of the functions of the outer membrane probably is to deter selectively the approach of many compounds. The LPS molecules cover the cell with a very hydrophilic surface. This should serve to impede the entry of any hydrophobic compound, which probably explains why gram-negative bacteria are relatively resistant to many hydrophobic antibiotics. (Penicillin G is hydrophobic. Its analogs or derivatives, such as ampicillin, are less hydrophobic and are consequently more effective against gram-negative bacteria.) On the other hand, hydrophilic compounds larger than 600 molecular weight are excluded by the molecular sieving functions of the matrix protein.

The periplasmic space In this compartment of gram-negative bacteria are located a number of proteins. They are water-soluble and usually quite stable. Generally, the proteins of the periplasm fall into either of two classes: degradative enzymes (for example, alkaline phosphatase and ribonuclease) or binding proteins for sugars, amino acids, or ions (for example, galactose-binding protein). These binding proteins serve as specific accessories of systems for active substrate transport into the cytoplasm and as signal receptors that orient bacterial swimming (see Chapter 3).

The osmotic pressure in the periplasm seems to be buffered by a set

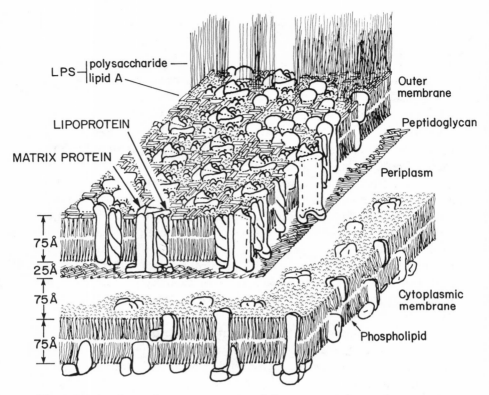

Fig. 2.14 A schematic representation of the gram-negative cell envelope. The polysaccharide chains on only a portion of the LPS molecules are shown. Note that phospholipid is relatively less abundant in the outer leaflet of the outer membrane. (Adapted from J. M. DiReienzo, K. Nakamura, and M. Inouye. 1978. *Annual Review of Biochemistry* 47:481–532.)

of membrane-derived oligosaccharides whose synthesis is increased when the environmental osmotic pressure becomes too high with respect to that in the cytoplasm. These molecules contain about 8–10 glucose units in a highly branched structure. On some of these glucose units glycerol or ethanolamine are attached by phosphodiester linkage. The complex molecules are just large enough so they cannot escape through the pores of the outer cell membrane, and can be numerous enough to augment significantly the osmotic pressure in the periplasm. Thus, thanks to their outer cell membrane, gram-negative bacteria are able to provide themselves with a *milieu intérieur*, like multicellular eukaryotes.

The Capsule

The outermost layer of both gram-positive and gram-negative cells is often a **capsule,** a loose gelatinous layer. This coat is usually quite thick, occupying a volume that is much larger than that of the cell, and can sometimes be seen by light microscopy of unfixed cells, using India ink as a negative stain. The capsular material often impedes phagocytosis (unless antibody to it is present) and sometimes plays a major role in virulence. Capsules are usually relatively simple polysaccharides or polypeptides, but protein capsules also exist. Some examples are:

Leuconostoc species	Dextran (glc-α-1,6-glc)$_n$
	Levan (fru-β-2,4-fru)$_n$
Bacillus anthracis	poly-D-glutamic acid
Streptococcus species	type 3 polysaccharide
	(glc-β-1,3-glucuronic-β-1,4)$_n$
	hyaluronic acid
	(glucuronic-β-1,3-glcNAc)$_n$

It is now recognized that in order for *B. anthracis* to be fully virulent, it must possess both the polyglutamic acid capsule and a protein exotoxin.

Capsules are important clinically because they are antigenic. For example, 83 antigenically different capsular types of *Streptococcus pneumoniae* are known. An antibody elicited by a given capsular antigen is specifically protective only against organisms with that antigen. When the antibody reacts with the capsular antigen, swelling of this layer occurs and the complex becomes refractile and visible in a light microscope; this is known as the *quellung* reaction.

Other Structures and Appendages

As shown in Figure 2.3, many bacteria contain *pili* (fimbriae); the ones illustrated are the sex pili, which are involved in bacterial conjugation and are found in small numbers (1–10) on the cell (see Chapter 7). In addition, many bacteria are surrounded by a mat of relatively short pili, which may be the principal antigen of the cell and sometimes act as the adhesive organelle in attachment to a

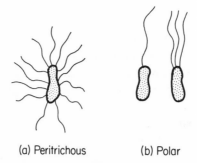

(a) Peritrichous   (b) Polar

Fig. 2.15 Distribution of bacterial flagella. (a) Peritrichous. (b) Polar.

particular tissue. Pili are composed of protein, in general formed by the polymerization of a single kind of subunit.

Some bacteria are motile; they swim by rotation of **flagella,** which are helical appendages originating in the cytoplasmic membrane. Flagella can be distributed all over the surface (peritrichously), as in *E. coli* and *Proteus,* or at one pole, as in *Pseudomonas* (Fig. 2.15). *Aquaspirillum magnetotacticum* has a flagellum at each pole. Although the flagellum proper is composed of a single polymerized protein (flagellin), the structure attaching the flagellum to the envelope is complex (see Chapter 3). Bacterial **chemotaxis** (movement toward attractive nutrients or away from toxic substances) involves the control of flagella rotation.

Questions

2.1. Treatment of gram-positive cells with agents interfering with peptidoglycan synthesis or structure gives rise to protoplasts. In protective conditions protoplasts sometimes can grow and replicate, even in the presence of the agent which caused their formation (for example, penicillin). (a) What are "protective conditions"? (b) Might they exist in the host? (c) Why are protoplasts insensitive to the drug which causes their formation? (d) In the absence of the drug, but still under protective conditions, certain protoplasts grow out as normal vegetative cells, but in other cases they remain as protoplasts. Why?

2.2. In gram-negative cells, alkaline phosphatase is located primarily in the periplasm. (a) Why is this enzyme not found in the cytoplasm? (b) In gram-positive cells, it is found extracellularly. Is that functionally reasonable? (c) How might the two different locations affect the role of the enzyme?

2.3. Gram-negative cells are relatively insensitive to penicillin G, but sensitivity is increased in a mutant lacking 0-antigen. Treatment of wild-type cells with the chelating agent ethylenediaminetetraacetate (EDTA), known to loosen or release LPS, also increases sensitivity to penicillin G. (a) How can the change in drug sensitivity be explained? (b) Would such treatment also increase sensitivity to actinomycin? (c) To lysozyme? (d) To cyanide?

2.4. Cell surface components are sometimes classified according to their reactions with antibody. Thus when the capsular material from a given strain of bacteria is used to immunize an animal, the antiserum elicited will contain specific antibodies against each kind of capsular polysaccharide present in that bacterial strain. This antiserum will agglutinate not only the organism whose capsule was used as antigen but also related strains that share any capsular component with that organism. The presence of different antigenic groups on the same organism can often be established by comparing the agglutinating activity of an immune serum before and after preadsorption with a second organism.

For example, suppose that intact cells of three clinical isolates of *Salmonella* X, Y, and Z are used separately to obtain antiserum. Each of these sera are then divided into four portions: one is not treated; one is mixed with an excess amount of strain X cells, and then the cells are removed by centrifugation; one is mixed with an excess amount of strain Y cells, and so on. The following list gives the results of a series of agglutination tests with the various kinds of sera:

Test	Antiserum to strain	Absorbed on strain	Agglutinated strain
a	X	None	X,Z
b	X	X	None
c	X	Y	X,Z
d	X	Z	X
e	Y	None	Y,Z
f	Y	X	Y,Z
g	Y	Y	None
h	Y	Z	Y
i	Z	None	X,Y,Z
j	Z	X	Y,Z
k	Z	Y	X,Z
l	Z	Z	None

Devise the simplest antigenic system that could account for these results, assuming that no masking or interference of one antigen by another is taking place. (Assign arbitrary numbers to each antigen(s): I, II, III, and so on.)

Strain X has antigen(s) ＿＿＿＿＿＿

Strain Y has antigen(s) ＿＿＿＿＿＿

Strain Z has antigen(s) ＿＿＿＿＿＿.

Note that in this example one is distinguishing different antigenic specificities, and not necessarily different molecules. Thus, the different antigens might be different antigenic determinants of the lipopolysaccharide, or determinants on different types of structure altogether (lipopolysaccharide, flagella, pili, and so forth).

2.5. Growing cells synthesize wall and septum at an equatorial growing point. This can be seen using short treatment with a radioactive material specifically labeling peptidoglycan (for example, ^{3}H-diaminopimelate [DAP], which is in position R3 of Figure 2.6). The freshly incorporated ^{3}H-DAP is seen to concentrate at the equatorial region. However, longer labeling with ^{3}H-DAP shows it to be evenly distributed all over the cell. What could be happening?

Further Reading

Inouye, M. 1979. Bacterial Outer Membranes: Biogenesis and Functions. New York: Wiley.

Leive, L., ed. 1973. Bacterial Membranes and Walls. New York: Marcel Dekker.

Mendoza, D. de, and J. E. Cronan, Jr. 1983. Thermal regulation of membrane lipid fluidity in bacteria. Trends in Biochemical Sciences 8:49–52.

Nikaido, H., and T. Nakae. 1979. The outer membrane of gram-negative bacteria. Advances in Microbial Physiology 20:163–250.

Osborn, M. J., and H. C. P. Wu. 1980. Proteins of the outer membrane of gram-negative bacteria. Annual Review of Microbiology 34:369–422.

Philip, J., and F. R. Whatley. 1975. Paracoccus denitrificans and the evolutionary origin of the mitochondrion. Nature 254:495–498.

Rogers, H. J., J. B. Ward, and I. D. J. Burdett. 1978. Structure and growth of the walls of gram-positive bacteria. In Relations between Structure and Function in the Prokaryotic Cell, ed. R. Y. Stanier, H. J. Rogers, and J. B. Ward. 28th Symposium of the Society for General Microbiology. Cambridge: Cambridge University Press.

Tipper, D. J., and A. Wright. 1979. The structure and biosynthesis of bacterial cell walls. In *The Bacteria: A Treatise on Structure and Function*, vol. 7, ed. J. R. Sokatch and N. N. Ornston. New York: Academic, pp. 291–426.

Ward, J. B. 1981. Teichoic and teichuronic acids: biosynthesis assembly, and location. *Microbiological Reviews* 45:211–243.

Wright, A., and D. J. Tipper. 1979. The outer membrane of gram-negative bacteria. In *The Bacteria: A Treatise on Structure and Function*, vol. 7, ed. J. R. Sokatch and N. N. Ornston. New York: Academic, pp. 427–485.

3

Bacterial Nutrition and Metabolism

ALL CELLS, including bacteria, are composed of a similar array of chemical elements which must be supplied by the environment: carbon, hydrogen, oxygen, phosphorus, and sulfur (found in organic compounds), and magnesium, iron, molybdenum, zinc, cobalt, potassium, and so on (found as ions). The proteins of all cells are made up of the same set of amino acids and the nucleic acids are composed of the same nucleotides; likewise, lipids and polysaccharides also have their typical basic units. But cells vary enormously in the ways they obtain these elemental building blocks and turn them into the organic molecules and energy necessary to sustain life.

In this chapter we will see that bacteria differ, first, in their minimal requirements for growth. Some species can grow on a very simple medium, whereas others require numerous growth factors. Second, organisms differ in the range of nutrient sources they can use. Versatile ones will grow on any one of many carbon and energy sources, whereas others need certain combinations. And third, bacteria show variety in the ways they derive energy from the nutrients they consume. All of these differences reflect the variations in biosynthetic and degradative pathways that have evolved in different organisms.

Minimal Requirements for Growth

Most bacteria of interest in medicine and genetics will grow on undefined rich media such as soups from boiled yeast or beef hearts, but the actual compounds needed for growth vary considerably from species to species. The precise growth requirements have been determined for many bacteria. Organisms such as *Escherichia, Bacillus, Pseudomonas,* and *Azotobacter* can grow in a rather simple medium. It must contain the usual inorganic ions (NH_4^+, $HPO_4^=$, $SO_4^=$, K^+, Mg^{++}, Fe^{++}, and so on) and at least a single organic compound such as glucose which can serve as a source of carbon and energy. At the other end of the range would be a bacterium such as *Streptococcus faecalis* which requires, in addition to a chief carbon and energy source, the following essential growth factors: isoleucine, leucine, valine, cystine, methionine, tryptophan, tyrosine, phenylalanine, alanine, aspartate, glutamate, proline, glycine, serine, threonine, arginine, lysine, histidine, adenine, guanine, uracil, pantothenic acid, riboflavin, thiamine, nicotinic acid, pyridoxine, p-aminobenzoic acid, biotin, and folic acid. If a single item from this list is omitted, there will be no growth of this organism. *S. faecalis* has been used for **bioassay** of many of these factors.

Since the final cell composition in all cases is similar, the difference in minimal requirements reflects differences in biosynthetic ability. Figure 3.1 outlines intermediary metabolism in an organism such as *E. coli* which has simple requirements. The metabolism of human cells would give a similar picture but with fewer available biosynthetic pathways, and *S. faecalis* would have even fewer. Thus, *E. coli* can build all the twenty amino acids, human cells can make certain "nonessential" ones (so we must eat the others), and *S. faecalis* can synthesize even fewer. Similarly, both *E. coli* and human cells can make nucleic-acid bases, but *S. faecalis* cannot. On the other hand, the three types of cells contain in common numerous other routes of intermediary metabolism, such as glycolysis, and use the same cofactors in the same enzymatic reactions. The absence of vitamins in the list of requirements of *E. coli* means that this organism has the biosynthetic capacity to make the cofactors (not shown in Fig. 3.1).

The generalization that bacterial nutritional requirements reflect biosynthetic inability must be qualified, however. In the history of microbiology there have been many findings of nutritional requirements that turned out to be spurious. For example, a peptide might be needed, while the constituent amino acids would not suffice. The reason may be that there is a special transport system for the peptide but not for the constituent amino acids or that freely available amino acids are used for metabolism in a wasteful manner and soon become

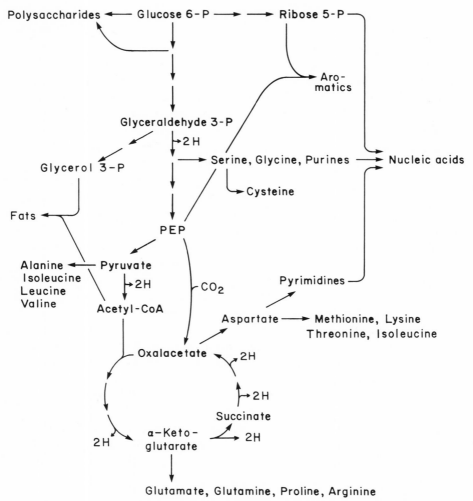

Fig. 3.1 The basic metabolic network of *E. coli.*

exhausted. Another case would be the apparent requirement of a particular protein; the tubercle bacillus was found to grow best with serum albumin, but the real function of the protein was that it sequestered traces of toxic fatty acids.

In the case of a "simple" amino acid requirement in an organism, such as streptococcus, it is not usually known whether the entire biosynthetic pathway is missing or just a single enzyme. A systematic analysis of the amino acid requirements of *Lactobacilli,* however, showed that some pathways suffered only minor genetic lesions, since mutants that dispense with the requirement of certain amino acids can be isolated. And finally there can be an apparent simple requirement which is actually a case of growth inhibition caused by other constituents in the medium. For example, some strains of *E. coli* are inhibited by valine, although they will grow in a minimal medium without it. The valine inhibition is prevented by isoleucine: that is, in the presence of valine there is an isoleucine requirement. This is a case of biosynthetic deficiency imposed by the medium.

CARBON DIOXIDE AS A BACTERIAL REQUIREMENT

Carbon dioxide (or bicarbonate) is needed for growth of all cells, prokaryotic and eukaryotic, for there are essential reactions of intermediary metabolism that use it. In the formation of dicarboxylic acids during growth on sugars, one of the carbon atoms of oxalacetate (and its derivatives) comes by carboxylation; and even in a rich medium, CO_2 is required for fatty acid biosynthesis.

General metabolism also produces CO_2, as in the decarboxylations of pyruvate and in the TCA cycle. In metabolism as a whole there is more decarboxylation than carboxylation occurring. In many cases CO_2 is supplied endogenously, therefore no exogenous requirement is apparent. However, the CO_2 and bicarbonate in the culture are in equilibrium with the CO_2 in the air. So even if bacteria do not show a CO_2 requirement when inoculated at high cell density, some do show it when streaked out as single cells on plates. For example, the organism that causes gonorrhea, *Neisseria gonorrhoeae*, will not grow on the richest medium if incubated in air (0.03% CO_2) but will grow well if the CO_2 concentration is increased to about 3%. This is easily arranged by allowing a candle to burn out in the jar for incubating plates.

Maximal Exploitation of Nutrient Sources

Just as bacteria differ in their minimal requirements for growth, they also differ in the number of compounds they are able to take up from the environment and use. Enteric bacteria such as *E. coli* have the ability to use many sugars. *Pseudomonas sp.* seem even more versa-

tile; some are able to use several hundred organic compounds—
sugars, organic acids, amino acids, complex heterocyclic organic
compounds—as sole carbon and energy sources. Such versatility is a
reflection of an organism's degradative (catabolic) pathways. They
may be short (as in the conversion of maltose to glucose) or quite long
(as in the conversion of glucuronic acid to pyruvate) and are generally inducible (see Chapter 5).

TRANSPORT SYSTEMS IN BACTERIA

The first step in a catabolic pathway is usually transport of the
specific nutrient into the cell. In some cases this step might be
preceded by hydrolytic reactions, because the nutrient is too bulky.
In order to utilize large molecules of cellulose, the gram-positive
Clostridium thermocellum has to secrete a group of cellulases that
catalyze the breakdown of the polymers to the disaccharide cello-
biose and to glucose, which are then transported into the cell. In
gram-negative cells, certain medium-sized molecules can pass
through the porin channels formed by the matrix protein of the outer
membrane (see Fig. 2.14) but still cannot be utilized because there
are no specific transport systems to carry them across the inner
(cytoplasmic) membrane. For example, a number of phosphoesters
that penetrate the outer membrane can only be used as a source of
phosphate or carbon after they are cleaved by periplasmic hydrolytic
enzymes. Movement of substances across the inner membrane is
accomplished by several kinds of mechanisms, which vary in the
energy they require.

Simple diffusion is a purely physical process that results in the
equilibration of the concentration of a compound in a space by
random molecular motion. Water might enter or leave the cytoplasm
in this manner.

Facilitated diffusion, like simple diffusion, requires no energy. So,
concentration of the compound against a gradient does not occur.
However, facilitated diffusion does require a specific membrane
protein.

Active transport is another carrier-mediated process and is therefore specific, saturable, and can be competitively inhibited. In addition, the substrate is accumulated against a concentration gradient
(sometimes as high as 10^5, inside/outside), which requires the expenditure of metabolic energy.

Group translocation requires energy but is not strictly an active-transport process because the compound is not concentrated unchanged. Group translocation requires a specific carrier which modifies the substrate chemically during translocation and then releases the product into the cytoplasm. Once the substrate is modified, for example by phosphorylation, the product can no longer pass the membrane barrier and is thus trapped inside the cell. In this process the metabolic energy is expended at the modification step.

CARBOHYDRATE TRANSPORT SYSEMS IN E. COLI

Some examples of the transport systems that E. coli possesses to transfer sugars into the cytoplasm are schematically depicted in Figure 3.2. Of the carbohydrates shown, only the smallest one, glycerol, enters the cell by facilitated diffusion. Once the compound enters, its escape is prevented by a phosphorylation reaction catalyzed by a cytoplasmic enzyme, glycerol kinase, with ATP as the phosphoryl donor.

Active transport systems Active transport requires the expenditure of metabolic energy. In both bacteria and higher cells, energy may be obtained from two sources: ATP and electrochemical gradients (often referred to as the proton motive force). In bacteria, as in mitochondria of higher cells, the two types of potential energy are interconvertible by the membrane ATPase: ATP hydrolysis can establish a membrane potential gradient (Fig. 3.3) and, vice versa, a potential gradient can drive ATP formation. Figure 3.4 is a schematic drawing of the E. coli membrane, including the steps whereby ATP is made (see also "Respiration," below).

E. coli exhibits a number of variations on active transport. Of the single component systems, the most extensively studied is that of lactose. This disaccharide can readily cross the outer membrane through the pore formed by the matrix (porin) protein, a process that does not require energy. The substrate is then recognized and carried into the cytoplasm by an integral protein of the cytoplasmic membrane. The substrate can be concentrated in unaltered form up to about 1,000-fold. This transport system functions as a **symport** (a mechanism requiring the simultaneous transport of two different substrates in the same direction) with H^+ as the cosubstrate (see Fig. 3.4). Since energized cells have fewer protons inside than outside,

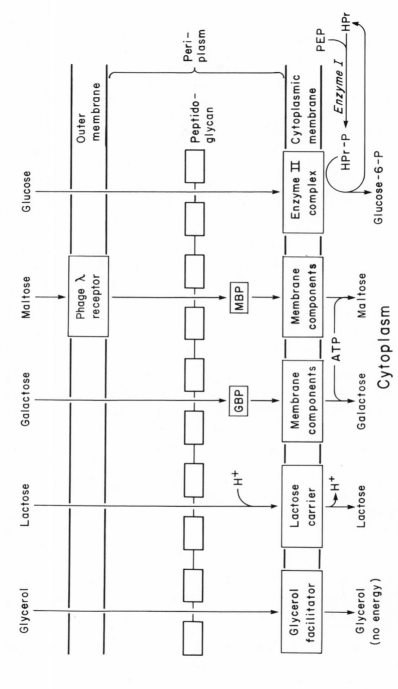

Fig. 3.2 A schematic representation of the various modes of carbohydrate transport by *E. coli*. (GBP = galactose-binding protein; MBP = maltose-binding protein; and PEP = phosphoenolpyruvate.)

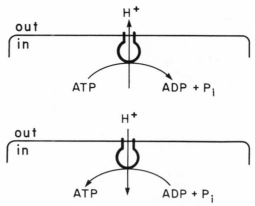

Fig. 3.3 The interconversion by membrane ATPase of electrochemical energy stored in ATP. This scheme and others describing bacterial energy metabolism are greatly simplified.

protons will be running down the gradient when they enter the cell. Lactose, which rides the same carrier, will be running against a gradient. Thus the energy source for this active system is the proton motive force.

An example of a more complex transport system in *E. coli* is the galactose pump, which contains three protein components. Two of these are associated with the cytoplasmic membrane and the other, the galactose-binding protein, is in the periplasmic space. The mechanism for powering this kind of transport is not yet known. However, the substrate is not phosphorylated during transport. The extra complexity seems useful for several reasons. First, systems with periplasmic proteins generally have 100 to 1,000 times greater affinity for the substrate than do transport systems without periplasmic binding proteins. Second, these systems can maintain concentration gradients as high as 10^5. In other words, transport systems of this nature have high extracting power. Third, the periplasmic protein serves as a signal receptor directing cell chemotactic movement.

The maltose transport system involves a specific protein structure in the outer membrane, the λ receptor (so called because the bacteriophage λ uses this protein as a receptor), as well as a periplasmic binding protein and an integral protein in the cytoplasmic membrane. Maltose is small enough to pass through the outer membrane by diffusion across the matrix protein pore. However, the sugar

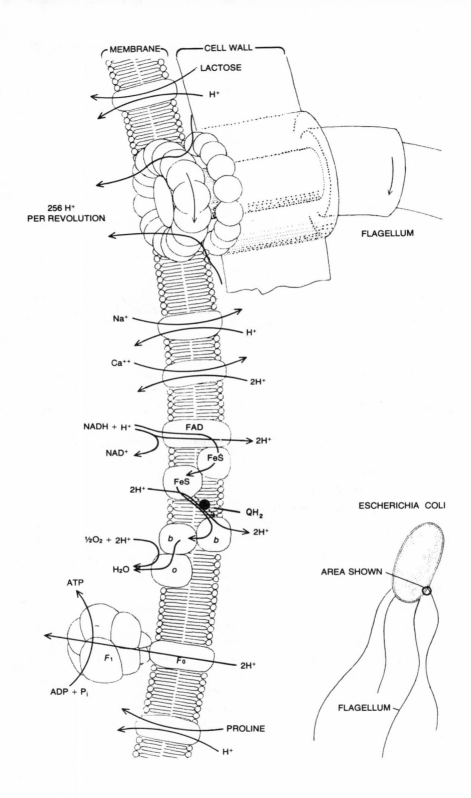

MEMBRANE

CELL WALL

LACTOSE

H^+

256 H^+
PER REVOLUTION

FLAGELLUM

Na^+

H^+

Ca^{++}

$2H^+$

$NADH + H^+$

FAD

$2H^+$

NAD^+

FeS

FeS

$2H^+$

QH_2

$2H^+$

$\frac{1}{2}O_2 + 2H^+$

b

b

H_2O

o

ESCHERICHIA COLI

ATP

AREA SHOWN

F_1

F_0

$2H^+$

$ADP + P_i$

FLAGELLUM

PROLINE

H^+

passes through the outer membrane more readily with the aid of the maltose-specific channel. Accordingly, this structure is required for maltose transport at very low external substrate concentrations. The maltose system also transports larger glucose polymers, the malto-dextrins. These molecules are too large to pass through the matrix

---

Fig. 3.4 Cytoplasmic membrane of the bacterium *E. coli:* a schematic cross-sectional view with the location of important membrane proteins. In the phospholipids (shaped like clothespins in this drawing) circles indicate the polar head groups and lines indicate fatty acid side chains. These phospholipids are arranged in the typical bilayer fashion. The membrane-associated respiratory chain is represented as bodies embedded in the phospholipids, starting at the middle with oxidation of NADH by a flavin adenine dinucleotide (FAD)-associated dehydrogenase. This reaction is accompanied by the expulsion of two H^+. (The NAD^+ produced can then participate in oxidizing a cytoplasmic substrate [SH_2] with the reformation of NADH, not shown in the figure.) The pair of electrons transferred to the FAD protein are returned to the inner surface through iron-sulfur proteins (FeS). These electrons, together with two protons from the cytoplasm, then reduce a molecule of ubiquinone to hydroquinone (QH_2, which is indicated by the filled circle). This carrier in turn passes the electrons and protons to cytochrome b, which extrudes the two H^+ into the environment and passes the two electrons to cytochrome oxidase (o), which finally reduces molecular oxygen, forming H_2O. For each molecule of NADH oxidized, four H^+ are translocated.

This translocation energizes the membrane by creating a proton gradient (proton motive force). Energy stored in this gradient can then be used to drive various cellular processes. For example, two H^+ running into the cytoplasm down the gradient through ATPase (F_1 and F_0 subunits) will drive the formation of one molecule of ATP. The reverse of this reaction will pump two H^+ out of the cytoplasm. The proton motive force can also drive transport of various sugars and amino acids into the cytoplasm or ions out of it. The examples shown here are the lactose/H^+ (top) and proline/H^+ (bottom) *symports* and the Na^+/H^+ and Ca^{++}/H^+ *antiports* (middle). Finally, the proton motive force can be utilized to drive flagella rotation.

Note the arrangement of the various proteins in the bilayer structure. The F_1 subunit of the ATPase is not embedded in the bilayer, but is just attached to the membrane. This is therefore a peripheral membrane protein. Various components of the respiratory chain are embedded in the bilayer (FeS, b, o). These are integral membrane proteins. The transport carrier molecules (ATPase-F_0, lactose/H^+, and so on) pass entirely through the bilayer structure and are therefore transmembranal proteins. The envelope layers of *E. coli* that are external to the cytoplasmic membrane are not shown in this diagram. (Adapted from P. C. Hinkel and R. E. McCarty. 1978. How cells make ATP. *Scientific American* 228:104–123. Copyright © 1978 by Scientific American, Inc. All rights reserved.)

protein pore and therefore exhibit an absolute requirement for the outer membrane receptor.

Group translocation systems   A number of different bacteria, including *E. coli*, transport a variety of sugars by a process that delivers sugar phosphates in the cytoplasm. These sugars are phosphorylated in transit by protein complexes referred to as the phosphotransferase systems (PTS). Such systems consist of three components: HPr (a histidine-containing protein), Enzyme I, and an Enzyme II complex. They are involved in the following reactions:

$$\text{P-enolpyruvate} + \text{HPr} \xrightleftharpoons[]{\substack{\text{Enzyme I} \\ \text{(cytoplasm)}}} \text{Pyruvate} + \text{P-HPr} \quad \text{(Reaction 1)}$$

$$\text{P-HPr} + \text{Sugar} \xrightarrow{\substack{\text{Enzyme II complex} \\ \text{(membrane)}}} \text{Sugar-P} + \text{HPr} \quad \text{(Reaction 2)}$$

$$\underset{\text{(outside)}}{\text{P-enolpyruvate} + \text{Sugar}} \rightarrow \underset{\text{(inside)}}{\text{Sugar-P}} + \text{Pyruvate} \quad \text{(Overall reaction)}$$

The first reaction occurs in the cytoplasm and is a preparative step in the transport of any PTS sugar. Note that the driving force of the translocation is the high-energy phosphate bond of phosphoenolpyruvate. The second reaction, the actual translocation step, occurs in the cytoplasmic membrane and is catalyzed by the Enzyme II complex. This complex is sugar-specific, that is, there are different Enzyme II complexes for different sugars. Once the sugar is inside and phosphorylated, it is trapped and cannot exit. The phosphorylation which occurs during the transport process is the first step in the metabolism of that sugar. This kind of process is therefore energetically economical.

The PTS system also has other important ancillary functions. For example, the engagement of the glucose Enzyme II by its substrate inhibits, through the action of a third protein (factor III^{Glc}), the activities of other transport proteins, such as the one for lactose or the one for glycerol. In addition, translocation of glucose causes a decrease in the cellular concentration of cyclic AMP, which in turn impedes the expression of certain genes involved in catabolic pathways (see Chapter 5).

VARIATIONS IN TRANSPORT SYSTEMS

In the discussion of $E.$ $coli$, various transport systems have been divided into classes according to the location of the proteins involved, the type of energy coupling, and the mechanism by which substrate uptake is accomplished. This is, however, in several ways a simplified view of transport in the bacterial world.

First, not all bacteria contain all of the different transport systems that $E.$ $coli$ has. Gram-positive bacteria, which do not have an outer membrane, are not equipped with transport systems analogous to the ones used by $E.$ $coli$ for galactose and maltose transport. Furthermore, most strictly aerobic bacteria transport nearly all sugars by active concentration rather than by the PTS-mediated translocation process.

It follows that the same sugars are not transported in all bacteria in the same way. The gram-positive $Staphylococcus$ $aureus$, for instance, transports lactose by the PTS but transports maltose via facilitated diffusion. The gram-negative $Pseudomonas$ $fluorescens$ concentrates maltose, but without a periplasmic protein.

Finally, a substance may be transported by different mechanisms in the same organism. For example, there are seven known ways by which $E.$ $coli$ transports the essential trace element iron. Since it is relatively insoluble even in ionic form, bacteria have developed very high affinity transport systems for this element. Many of these involve iron-binding chelates (siderophores) that are secreted into the growth media by the bacteria. One such chelate, enterochelin (Fig. 3.5), is synthesized by $E.$ $coli$ and secreted into the growth media. Enterochelin binds the metal ion and the Fe^{3+} complex is taken by an active transport system. The enterochelin-Fe^{3+} complex is too large to pass through the pore of the outer membrane. It is not surprising, therefore, that at least one (and possibly two) outer membrane proteins must intervene.

Enterochelin binds Fe^{3+} so tightly that in order to release the iron into the cytoplasm the chelating agent must first be degraded. This is accomplished by a specific esterase. The Fe^{3+} is then reduced to Fe^{2+} prior to utilization.

Besides utilizing enterochelin, $E.$ $coli$ can exploit chelating agents that are synthesized and secreted by other microorganisms. This undoubtedly provides the organism with a great selective advantage under conditions of iron-limited growth. For example, $E.$ $coli$ can also

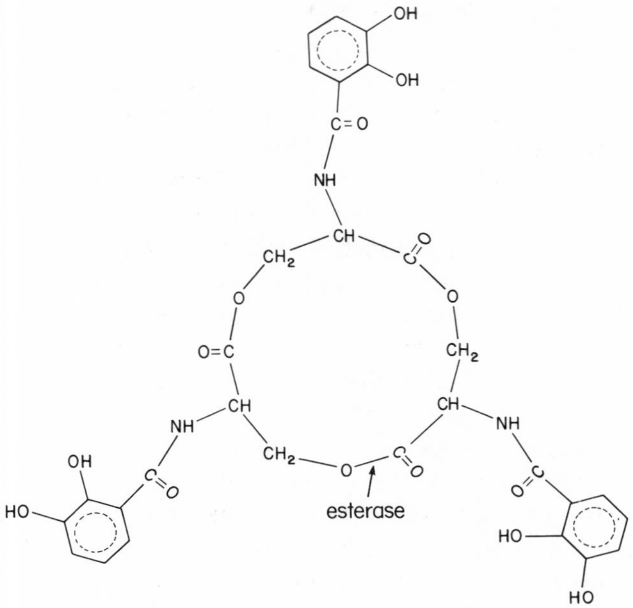

Fig. 3.5 Structure of enterochelin.

transport iron chelated by the fungal siderophores ferrichrome and rhodotorulic acid.

CHEMOTAXIS

We saw in Chapter 2 that many bacteria with flagella have the "sense" to swim toward higher concentrations of nutrients (attractants) or away from harmful substances (repellents). This phenomenon, called **chemotaxis,** is illustrated in Figure 3.6. An inoculum of bacteria was deposited in the center of a soft agar plate of high water content (so the bacteria can swim) containing a growth-limiting concentration of the attractant galactose. As the cells began to grow, the galactose was used up, creating a concentration gradient. The bacteria, sensing this gradient, swam away from the center so that after overnight incubation, they have formed a ring surrounding the depleted medium.

Only some nutrients are attractants. Gluconic acid, another carbon source for *E. coli* which supports the same growth rate as galactose,

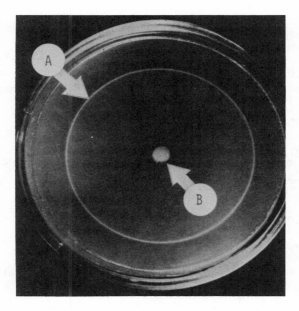

Fig. 3.6 Swarming of *E. coli* cells away from the center of an agar plate in search of galactose. (A) Visible ring of *E. coli* cells migrating away from the center; (B) the spot where the initial inoculum was placed. (From J. Adler. 1966. *Science* 153:708–716. Copyright 1966 by the American Association for the Advancement of Science.)

does not attract the cells. In the analogous experiment with gluconic acid, the cells would still be at the center of the plate after overnight incubation, although the spot size would have increased because of random swimming and population growth. Likewise, in the galactose case, a nonchemotactic mutant would remain at the center.

The cumulative direction of swimming is controlled by the timing of changes of flagella rotation. *E. coli*, which is peritrichous, has several flagella. When activated by an attractant, all of the flagella rotate counterclockwise in a coordinated manner, forming a tuft that propels the cell forward linearly. When the flagella reverse the direction of rotation, coordination is lost and the tuft flies apart; swimming stops and the cell "tumbles." The flagellar apparatus is the only known kind of biological structure that functions like a rotor (Fig. 3.7). More than 30 genes are known to be involved in the orderly synthesis and functioning of the apparatus.

In a homogeneous medium without a gradient of attractant or

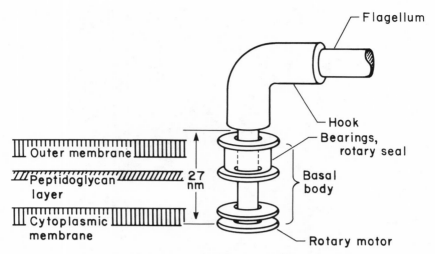

Fig. 3.7 A model of the flagellar base. (Adapted from M. L. DePamphilis and J. Adler. 1971. *Journal of Bacteriology* 105:384–395.)

repellent, the bacteria will typically swim in a straight line for a few seconds and then tumble for a second or two before swimming in a new direction at random. When there is a gradient of an attractant, however, the duration of straight-line swimming will depend on the direction: cells swimming up the gradient will continue this activity for a longer period of time before tumbling than cells going down the gradient; the reverse occurs when there is a gradient of a repellent. The net result will be a massive population migration in the same general direction by "biased random walk." What the cell senses is not the absolute concentration of a chemical but a change in its concentration. A few seconds to a few minutes after such a change is perceived, the cell no longer remains in the stimulated state and returns to the usual swim and tumble schedule. This reduced sensitivity reflects **accommodation,** a phenomenon encountered frequently in other sensory systems in biology. The main chain of events is now understood in outline.

First, the stimulant is recognized by a receptor. In *E. coli*, a periplasmic binding protein or the Enzyme II component of the PEP-phosphotransferase system serves as a specific receptor for a sugar attractant. Cytochrome o, which bridges the cytoplasmic membrane, serves as the receptor for molecular oxygen. Over 25 chemoreceptors have been identified in this bacterium. However, some agents can modulate the cell behavior not through the intervention of specific receptors but by affecting the proton motive force. Thus, *E. coli* seems to sense a drop in pH in the physiological range as a repellent.

Second, after the stimulant has combined with the **receptor protein,** this complex interacts with a group-specific **signaler protein** (MCP or methyl-accepting chemotaxis protein) which integrates the incoming information. For example, both the galactose- and ribose-binding proteins interact with a common signaler protein, MCP I, whereas the maltose-binding protein interacts with MCP II.

Third, the relay action of the signaler protein depends on the degree of methylation of its glutamyl carboxyl groups. Initially a positive signal from the receptor protein causes the signaler protein to transmit the message to the tumble regulator complex. The result is the release from the complex a CheZ **regulator protein.** CheZ then combines with the FlaB **motor protein,** causing the counterclockwise rotation of the flagella to be maintained.

Finally, a positive signal also triggers progressive enzymic methylation of the signaler protein by S-adenosylmethionine, thereby reducing the activity of the protein. When the methylation reaches a plateau, full accommodation is achieved; the CheY regulator protein is released from the tumble regulator complex and combines with the FlaA motor protein, causing the flagella to turn clockwise and the cell to tumble. Thus the accommodated cell and the nonstimulated cell behave in a similar way: both swimming and tumbling occur periodically at the routine schedule.

An increasing concentration of an attractant or a decreasing concentration of a repellent constitutes a positive signal. The opposite conditions constitute a negative signal. A negative signal promotes cell tumbling and demethylation of the signaler protein. Thus, the positive signal is accompanied by desensitization and the negative signal is accompanied by resensitization.

Energy Production

There are two forms of energy currency in both bacteria and higher cells: ATP and electrochemical gradients (the proton motive force). ATP drives many biosynthetic reactions, and electrochemical gradients drive other functions like chemotaxis and certain substrate transport systems. ATP is formed directly from fermentation catalyzed by cytoplasmic enzymes (a set of reactions not requiring exogenous hydrogen acceptors) and the proton motive force is derived from respiration catalyzed by proteins associated with the plasma membrane (a set of electron transport reactions that does require exogenous hydrogen acceptors). These two types of potential energy are interconvertible by the membrane ATPase.

FERMENTATION

Energetically, the simplest types of bacteria are those that do not have organized electron transport systems linking substrate oxidation with O_2 reduction. Such cells largely depend on **substrate-level phosphorylations** for ATP formation. There are not many reactions of this kind in the metabolic network of organisms. The most familiar is the sequence in glycolysis:

$$\text{Glyceraldehyde-3-P} \xrightarrow[-2H]{+P_i} \text{1,3-diphosphoglycerate} \xrightarrow{+ADP}$$

$$\text{glycerate-3-P} + \text{ATP}.$$

Those bacteria that do not have conventional electron-transport-driven phosphorylation, such as *Clostridium* and *Streptococcus*, depend on the membrane ATPase to form electrochemical gradients. Since during anaerobic growth NAD^+ has to be regenerated from NADH, a hydrogen acceptor has to be made available. In bacteria such a compound is usually an endogenous organic metabolite which is reduced and excreted into the medium as a fermentation product. One simple example would be the formation of lactate by reduction of pyruvate (as in anaerobic muscle). This happens in some species of *Streptococcus*:

$$\text{Glucose} + 2\ \text{ADP} + 2\ P_i \rightarrow 2\ \text{lactic acid} + 2\ \text{ATP}.$$

In this particular example, since streptococci have almost no biosynthetic capacity to make the building blocks for macromolecules, lactic acid may be formed stoichiometrically from glucose in growth, with the organic material for cell growth being provided by the other nutrients in the medium.

Lactic acid production is the simplest mode of fermentation, but there are many others. Most bacterial sugar fermentations can be represented as variations on ways of using pyruvate as electron acceptor (Fig. 3.8). Glucose fermentation by some streptococci gives

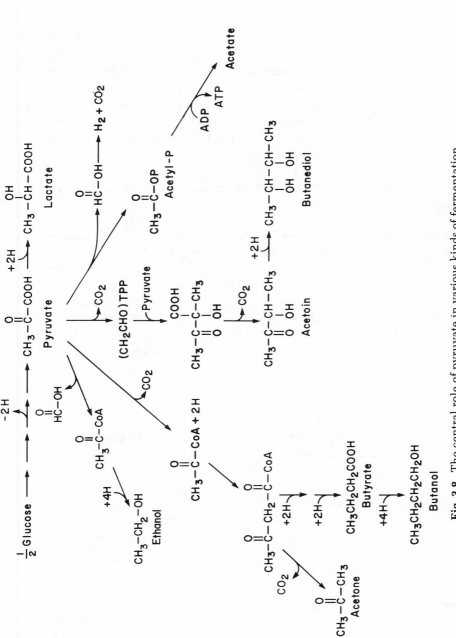

Fig. 3.8 The central role of pyruvate in various kinds of fermentation.

lactic acid, whereas *Clostridium acetobutylicum* makes acetone and butanol.

Enteric bacteria like *E. coli* can convert pyruvate to acetyl CoA and formate. Acetyl CoA can dispose of 4 H by being reduced to ethanol, or else acetyl CoA can be converted to acetyl phosphate which can in turn transphosphorylate with ADP, giving ATP and acetate. (Acetyl CoA, of course, can also be used for biosynthesis.) Acetate is a characteristic product of the "mixed-acid" fermentation, but formate and acetate cannot be the sole products from sugars, because the NADH formed in glycolysis must be converted back to NAD^+. Therefore, some acetyl CoA is reduced to ethanol via acetaldehyde; or else pyruvate is reduced to lactate. (*E. coli* cannot make acetone, butanol, or butanediol.)

Some bacteria (such as *Klebsiella*) with general metabolism similar to that of *E. coli* produce more neutral fermentation products, using a pathway where "active acetaldehyde" condenses "head to head" with pyruvate and gives acetoin by decarboxylation. Most of the acetoin is then reduced to butanediol (butylene glycol), but the trace amounts of acetoin are used in the Voges Proskauer test to diagnose this fermentation. Thus the particular pattern of products is characteristic of the particular organism, and gas-liquid chromatography of culture medium often is used to identify the organisms grown.

RESPIRATION (OXIDATIVE PHOSPHORYLATION)

Many bacterial species, like mitochondria of higher organisms, can generate ATP through electron transport with molecular oxygen as acceptor — a process called **oxidative phosphorylation.** The electron transport system is found in the cytoplasmic membrane, and the transport of electrons results in the extrusion of protons into the exterior. The electrochemical gradient thus established can drive ATP formation via the ATPase, as illustrated topographically in Figure 3.4 and schematically in Figure 3.9.

The combined function of the electron transport system and the ATPase in the same membrane allows considerably more ATP formation per molecule of substrate than can be provided by fermentation. Furthermore, with oxygen serving as final electron acceptor, there need be no accumulation of toxic fermentation products. Thus,

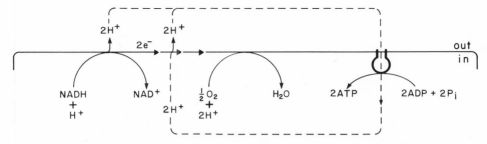

Fig. 3.9 The generation of ATP from ADP coupled to the oxidation of NADH.

aerobic growth gives more cells for the same amount of carbon source than anaerobic growth.

SPECIAL TYPES OF ELECTRON TRANSPORTS (ANAEROBIC RESPIRATION)

Oxidative phosphorylation is the most efficient way of generating metabolic energy by electron transport, but it is not the only way. The following processes are also widely distributed in nature:

Electron transport with organic compounds as acceptors Although anaerobes do not carry out conventional oxidative phosphorylation, a rudimentary form of electron transport phosphorylation is quite common among them. Thus, the oxidation of glycerol-3-P to dihydroxyacetone-P is often linked to the reduction of fumarate to succinate. Both reactions are catalyzed by membrane-bound flavoproteins, and their coupled function results in proton extrusion and generation of a proton motive force under anaerobic conditions in anaerobes as well as in facultative aerobes. The resulting electrochemical gradient can in turn generate ATP (Fig. 3.10).

Electron transport with nitrate as acceptor Bacteria with complex membrane electron transport systems often can use inorganic compounds, such as nitrate, as electron acceptor. A specifically induced enzyme complex, nitrate reductase, is the terminal component of such a chain (see "The Metabolism of Inorganic Nitrogen,"

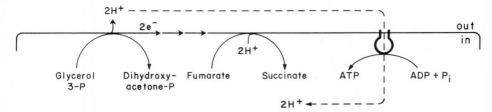

Fig. 3.10 The generation of ATP from ADP coupled to the conversion of glycerol 3-phosphate to dihydroxyacetone phosphate in the absence of air.

below). Because nitrate is a stronger oxidant than fumarate, this allows increased efficiency of electron transport (more protons expelled into the medium per pair of electrons removed from the primary substrate) in anaerobic situations (Fig. 3.11).

Energy from oxidation of inorganic substances Some bacteria have the ability to derive energy by the oxidation of inorganic substances, such as H_2 or $S^=$. The mechanisms used generally involve conventional electron transport systems leading to O_2, as acceptor, with associated phosphorylations. There are also bacteria that oxidize inorganic substances using nitrate as acceptor, and thus can grow on hydrogen gas at the expense of nitrate reduction.

Bacterial photosynthesis and autotrophic growth Bacteria also exist that use light as an energy source. The cyanobacteria (formerly called blue-green algae) perform a plant-like photosynthesis; another class performs strict anaerobic photosynthesis, using organic compounds or H_2 as reductant and not producing oxygen. Bacteria using

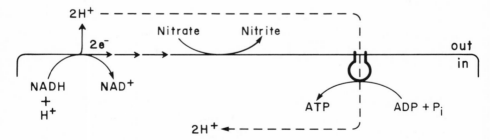

Fig. 3.11 The generation of ATP from ADP coupled to the reduction of nitrate in the absence of air.

light or oxidation of inorganic compounds as energy source usually can also grow with CO_2 as sole carbon source. This is called *autotrophy*.

Classification by Response to Oxygen

Bacteriologists frequently classify an organism according to whether oxygen is necessary, useful, useless, or actually toxic (lethal) to the cell. This is unfortunate, since a classification of bacteria based on the mode of energy production would more accurately reflect the evolutionary status of the organism. (For instance, the most primitive bacteria probably relied solely on fermentation. This was likely to be followed in sequence by the use of fumarate, nitrate, and molecular oxygen as the terminal electron acceptor for respiration.) Nevertheless, the following four categories are widely used in the literature as a convenient and practical way to describe bacteria.

Anaerobes indifferent to air (aerotolerant) A typical example would be the genus *Streptococcus*, which generally depends on substrate-level phosphorylation for ATP formation and cannot pass electrons to O_2 via membrane electron transport with proton extrusion. ATP formation and growth are thus relatively indifferent to air. However, such organisms do generally have soluble (instead of membrane-bound) flavoprotein oxidases (for example, $NADH + H^+ + O_2 \rightarrow H_2O_2 + NAD^+$). Such a reaction does not generate ATP.

Strict anaerobes In distinction to anaerobes that are indifferent to air, there are many organisms, such as *Clostridium*, that can grow *only* in the absence of oxygen, because oxygen is toxic to them. There are two classes of reaction using O_2 as final electron acceptor:

(1) $4e^-$ reductions giving water, as in electron transport: $4e^- + 4H^+ + O_2 \rightarrow 2H_2O$.
(2) $2e^-$ reductions giving hydrogen peroxide, as with flavoprotein oxidases: $2e^- + 2H^+ + O_2 \rightarrow H_2O_2$.

H_2O_2 can be toxic as an oxidizing agent and is usually eliminated by peroxidases: $H_2O_2 + H_2R \rightarrow 2H_2O + R$. (Catalase is a peroxidase catalyzing a reaction in which H_2O_2 itself acts also as the H_2R.)

Both classes of reaction that reduce O_2 create, as an intermediate or by-product, the O_2^- radical, called superoxide. Superoxide is a highly active free radical, toxic in many ways. Almost all bacteria and higher cells contain a detoxifying enzyme, superoxide dismutase, which catalyzes the reaction: $2O_2^- + 2H^+ \rightarrow H_2O_2 + O_2$. Strict anaerobes lack superoxide dismutase. Although they lack organized electron transport to oxygen, they still contain flavoprotein oxidases. The superoxide inevitably formed by these enzymes in the presence of O_2 cannot be destroyed.

The term "strict anaerobe" does not necessarily imply simple energy metabolism. The group includes the photosynthetic bacteria that grow on H_2S and contain electron transport systems as exotic as those found in chloroplasts, and the methane producers, whose energy metabolism is still not understood but which almost certainly perform electron transport phosphorylations.

Studies of strict anaerobes are cumbersome, because they have to be grown under conditions excluding O_2. This may be accomplished by adding reductants (often thioglycollate) to liquid media or by incubating plates in anaerobic jars (in which H_2 is generated chemically to consume the O_2 in the air). The majority of flora in the human large bowel are the strict anaerobe *Bacteroides*. Because of the practical difficulties of culturing this organism, less is known about it than one might suppose. Furthermore, the very enumeration of strict anaerobes in a specimen such as stool may be difficult, since their exposure to O_2 between the time of obtaining the specimen and plating it may be lethal. This is one example of a very general problem in medical bacteriology—that the composition of samples can change very quickly, depending on how they are handled before the analysis. A minor contaminant in a urine sample—minor because the bladder content is normally being diluted—may grow into a saturated culture just sitting overnight in a container. Various holding media have been devised to avoid such artifacts.

Facultative species Such bacteria can grow either aerobically, using electron transport phosphorylation, or anaerobically by fermentation. Probably most bacterial species belong to this class. Examples are *E. coli* and *Staphylococcus*.

Strict aerobes Some bacteria grow only aerobically. Examples are *Mycobacteria* (including the species that causes tuberculosis) and most species of *Pseudomonas*. The reason why an organism is an obligate aerobe is generally not known. Many have patterns of sugar metabolism somewhat different from those of facultative aerobes; strict aerobes lack a complete glycolytic pathway and metabolize sugars via oxidation to 6-phosphogluconate, which is then split (the Entner-Duodoroff pathway). They produce fewer acidic products from sugars and have electron transport pathways able to reduce certain aromatic amines ("oxidase positive" in diagnostic bacteriology). They probably all perform electron transport phosphorylation with O_2 as acceptor.

The Metabolism of Inorganic Nitrogen

In animal cells reactions of nitrogen are all at the level of ammonia (as in $NH_4^+ + 2H + \alpha$-ketoglutarate $\rightarrow$ glutamate), that is, there is no oxidation or reduction of the nitrogen. Many bacteria and plants can also obtain their nitrogen from nitrate ($NO_3^- \rightarrow NH_3$) in a process called **assimilatory nitrate reduction.** (This differs from dissimilatory nitrate reduction, or nitrate respiration, which leads to NO_2^-, NO_2, and N_2.)

Nitrate is in turn formed from NH_3 by a very specialized group of soil bacteria; the reaction provides energy via oxidative phosphorylation. This process is called **nitrification** (Fig. 3.12) and it conserves N in the soil by converting the volatile NH_3 to the nonvolatile NO_3^-.

Bacterial denitrification to give N_2, as well as leaching from the soil, leads to the loss of nitrogen in forms useful to plants (NH_4^+ and NO_3^-). Replenishment of NH_4^+ by reduction of atmospheric nitrogen (N_2) to NH_3 is called **nitrogen fixation.** The ability to carry out this

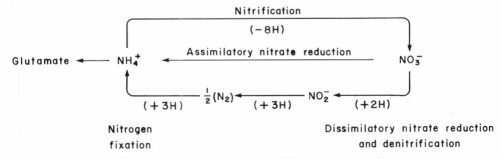

Fig. 3.12 A simple scheme of the nitrogen cycle.

fixation is fairly widespread among bacteria. Eukaryotes can carry out this reaction only in cooperation with prokaryotes. An important group of bacteria that performs this process is called *Rhizobium*, which lives in close association with leguminous plants. Thus, legumes have no need for nitrogenous fertilizers. Because the chemical synthesis of such fertilizers (for example, by the Haber process of N_2 reduction) is energy intensive and therefore expensive, there is considerable effort by biologists to improve our understanding of biological nitrogen fixation and extend its range.

Why Study Bacterial Nutrition?

Apart from their usefulness in taxonomy, studies on specific bacterial nutritional requirements and diversity have been carried out for a number of other reasons. There are still microbes of importance whose nutritional requirements appear to be so complex that the cells cannot be cultivated *in vitro* even in rich media. *Treponema pallidum*, which causes syphilis, is an example. Organisms like *Rickettsia* and *Chlamydia* might have even more fastidious nutritional requirements, for they grow only inside host cells. Culturing bacterial pathogens in vitro is invaluable to the understanding of how bacteria parasitize their hosts. In addition, many normal constituents of higher cells were first identified as bacterial growth factors (methionine is an example), and new growth factors are still being found in studies of bacterial nutrition.

Not all pathogenic bacteria, however, are nutritionally fastidious. *Staphylococcus aureus*, which can cause boils and carbuncles, grows in mineral media containing glucose, amino acids, thiamine, and nicotinic acid. *Cholera vibrio*, *Salmonella typhimurim* (causative agent of acute gastroenteritis or "food poisoning"), and *Shigella dysenteriae* have even simpler growth requirements: they can be cultivated in glucose-mineral media.

Nutritional requirements and diversity reflect the ecological niches of bacteria. The fastidious requirements of organisms like *Rickettsia* and *Chlamydia* are associated with their genetic adaptation to an intracellular life. Genes responsible for many biosynthetic pathways were no longer protected from deletions and other mutations by natural selection, because the biochemical products were supplied by the host cells. The common habitats of *Staphylococcus aureus* are the surfaces of nasal and mucous membrane and skin (hair follicles). Hence it should not be surprising that this organism can grow on glucose only when several amino acids and vitamins are also present; these minor nutrients are readily available in external biological fluids, including sweat. In contrast, enteric organisms do not take up permanent residence in their hosts. A large part of their populations must survive and propagate in the sewage and polluted waters where the presence of no particular metabolite in sufficient concentration can be counted upon. The ability to grow on a wide range of metabolites, and to synthesize all of the biochemical building blocks when necessary, therefore confers strong selective advantage.

Pathogenic bacteria are not always passively dependent upon the rich nutrients in the tissue fluid of their hosts. Many bacterial species can produce proteins that kill or injure host cells, thus liberating their nutritional contents. An example is diphtheria toxin, whose synthesis and secretion by the pathogen is stimulated by iron deficiency. The entry of one toxin molecule into a mammalian cell is sufficient to kill. An example of a less virulent biochemical strategy is the production by *Streptococcus mutans* (causative agent of dental caries) of a polysaccharide from sucrose, which allows the bacteria to adhere to the tooth surface, thus assuring their residence in a nutritious environment.

The need to acquire more knowledge about bacterial metabolism is not confined to organisms of clinical interest. An increasing number of bacterial species are being exploited for the production of chemical compounds from renewable sources such as starch and cellulose. The products include amino acids, proteins for livestock, and small organic compounds for the synthesis of commercially important polymers. Special strains of bacteria are also being developed for the degradation of chemical wastes. Effective use of such organisms will depend on knowing their physiological requirements and on understanding the biochemical pathways involved.

Questions

3.1. Bacteria contain many active transport systems for small molecules such as carbohydrates, but in general most animal cells obtain these compounds by facilitated diffusion. Why the difference?

3.2. *E. coli* mutants are known in which the lactose carrier protein is altered so that lactose transport occurs only by facilitated diffusion; there is no coupling to energy and no concentration of the substrate against a gradient. In what situation would this make a difference to growth on lactose?

3.3. There exist mutants of *E. coli* that do not make outer membrane pore. Are there any conditions where growth of such a mutant would be normal?

3.4. Active substrate uptake depends on energy. For *Streptococcus lactis,* uptake of radioactive TMG (a nonmetabolizable analog of lactose) occurs only in the presence of a metabolizable carbon source, such as glucose. The uptake of TMG is abolished by the ionophore carbonyl-cyanide-chlorophenyl-hydrazone (CCCP), even though this treatment does not immediately reduce the ATP level in the cell. Chromatography shows that the accumulated TMG is chemically unchanged. But if the same experiment is done with certain other streptococci, CCCP does not block accumulation, and the accumulated TMG is found as the phosphorylated compound. How can the different observations be explained?

3.5. In Figure 3.6, why did the descendants of some wild-type cells remain at the center and not join the swarming ring of cells?

3.6. *Leuconostoc mesenteroides,* which does not perform electron transport phosphorylation, has a pathway of glucose metabolism.

$$\text{Glucose} \rightarrow \text{glucose-6-P} \xrightarrow{-2H} \text{6-P-gluconate} \xrightarrow{-2H, -CO_2}$$
$$\text{pentose-P} \xrightarrow{+P_i} \text{acetyl-P} + \text{glyceraldehyde-3-P.}$$

Anaerobically the final products are lactate + ethanol. But aerobically the final products are lactate + acetate. Furthermore, the yield of cells/g glucose, a measure of the ATP made, is higher aerobically than anaerobically. Explain the different outcomes.

3.7. Can there be an organism without superoxide dismutase that is not a strict anaerobe?

Further Reading

Ames, G. F.-L., and C. F. Higgins. 1983. The organization, mechanism of action, and evolution of periplasmic transport systems. *Trends in Biochemical Sciences* 8:97–100.

Dills, S. S., A. Apperson, M. R. Schmidt, and M. H. Saier, Jr. 1980. Carbohydrate transport in bacteria. *Microbiological Reviews* 44:385–418.

Gottschalk, G. 1979. *Bacterial Metabolism.* New York: Springer-Verlag.

Haddock, B. A., and W. A. Hamilton. 1977. *Microbial Energetics.* Cambridge: Cambridge University Press.

Harold, F. M. 1972. Conservation and transformation of energy by bacterial membranes. *Bacteriological Reviews* 36:172–230.

———— 1978. Vectorial metabolism. In *The Bacteria: A Treatise on Structure and Function,* vol. 6, ed. L. N. Ornston and J. R. Sokatch. New York: Academic, pp. 463–513.

Hinkel, P. C., and R. E. McCarty. 1978. How cells make ATP. *Scientific American* 238:104–123.

Koshland, D. E., Jr. 1981. Biochemistry of sensing and adaptation in a simple bacterial system. *Annual Review of Biochemistry* 50:765–782.

Maloney, P. C., E. R. Kashket, and T. H. Wilson. 1974. A protonmotive force drives ATP synthesis in bacteria. *Proceedings of the National Academy of Sciences, USA* 71:3896–3900.

Morris, J. G. 1975. The physiology of obligate anaerobiosis. *Advances in Bacterial Physiology* 12:169–246.

Postgate, J. R. 1982. *The Fundamentals of Nitrogen Fixation.* Cambridge: Cambridge University Press.

Thauer, R. K., K. Jungermann, and K. Decker. 1977. Energy conservation in chemotropic anaerobic bacteria. *Bacteriological Reviews* 41:100–180.

4

Growth of Bacterial Cultures

IN THE LABORATORY, bacteria are grown on the surface of a nutrient gel or in a liquid medium as a suspension of cells. The material in the medium is usually sterilized by heating in a steam autoclave. Heat-labile material is sterilized by filtration under aseptic conditions and the sterile ingredient is added separately. Solid medium is commonly prepared by adding powdered agar (a polysaccharide derived from algae) in 1–2% concentration (that is, 1–2 g/100 ml) to a liquid medium. The agar dissolves during the steam sterilization; the mixed content is poured into Petri dishes of 9 cm diameter. Upon cooling, the agar causes the entire medium to gel, forming a plate of nutrient medium.

Growth on Solid Media

Bacteria are deposited on agar plates by streaking with a wire loop dipped into a cell suspension or by pipetting and spreading small amounts of a dilute cell suspension over the surface. This process is called **inoculation.** The Petri dishes are then incubated (usually at 37°C in medical bacteriology). Most bacteria cannot degrade the agar but they are able to feed on other components of the medium. If it is nutritionally adequate, single cells will form piles of cells (a **colony**) by repeated growth and division of the original cell. The progeny of a single cell grown asexually is called a **clone.** (When two cells happen to have been deposited in the same place on the plate by chance or, more commonly, when cells in the original sample were aggregated, the piles of cells are not considered a clone. *Staphylococcus,* for example, often grows in aggregates and some species of *Streptococcus* grows in chains.)

It follows that all cells in a clone are genetically identical, except for the inevitable mutants that make up a low proportion of the colony. A "pure" culture can be maintained by repeated subculturing of a colony, by restreaking from a colony to another plate, or by aseptic transfer of cells from a colony into liquid medium. For most biochemical and genetic work, pure cultures are necessary.

Over the whole bacterial world there are wide variations in the rate of growth, and hence in the time it takes to make a visible colony. Many bacteria of medical interest, as well as bacteria used in genetic studies, can double in less than 30 minutes when they are in favorable environments, and colonies appear within 10–24 hours of incubation.

Growth on solid medium is often used simply to determine the number of cells alive in a liquid suspension or culture. For this purpose a portion of a total sample is appropriately diluted and a measured volume is then spread evenly on the agar surface (plating). Following an appropriate incubation period, the colonies are counted: this is known as a **viable count.**

SELECTIVE AND DIAGNOSTIC MEDIA

In medical bacteriology the original specimen is often a mixture of many types of bacteria (as in stool or sputum). Even in disease, most of these bacteria are of no particular interest—they are the "normal flora." Therefore, the first step in clinical bacterial identification is usually to separate the various bacteria in a specimen by streaking on agar plates to give isolated colonies. Different bacteria give colonies of somewhat different appearance: size, color, texture, mucoidy, shape, and other properties. Identifications of bacteria are not made primarily on the basis of colony appearance, but it does allow one to pick the likely candidates for further tests.

One's ability to clone a particular bacterium from a mixed sample is greatly improved by the proper choice of medium. Rich media in medical bacteriology come in many varieties, with names like nutrient broth, Mueller-Hinton blood plates, and so on, which, usually for unknown reasons, favor the growth of particular bacteria. Blood agar plates contain, in addition to the routine ingredients, 5% sheep or horse blood. The purpose is to render visible the production of hemolysins (toxic proteins) by the colonies; zones of agar under these colonies are without intact red cells and appear brownish red, because of the denaturation of the liberated hemoglobin.

The medium can be selective, so that certain organisms — which may be in the majority — will not grow. Thus, to find *E. coli* in a mixture of *E. coli* and *Streptococcus faecalis* one could use a simple minimal medium without growth factors. Thayer-Martin medium for isolation of gonococcus from rectal samples is a "chocolate" agar plate supplemented with several antibiotics to which the gonococcus is naturally resistant. Chocolate agar contains blood that has been heated; this releases some extra metabolites (for example, NAD) and prevents their enzymic breakdown. Chocolate agar does not show hemolysis but is very rich.

To assess stool flora, MacConkey lactose agar — an "indicator" medium which contains bile salts and crystal violet to inhibit growth of gram-positive bacteria — is often employed. Rare lactose-negative cells (usually not *E. coli*) may be found on these primary isolation plates as pale colored (instead of red) colonies. Should they be too few in the initial sample, preliminary liquid enrichment media are used to select against the majority lactose-positive cells (for example, in selenite enrichment broth) before plating.

Most media in diagnostic bacteriology are poorly buffered so as to allow the use of acid-base indicators to show fermentations. Most bacteria do not grow when the pH falls below 5. Yeasts, however, do grow at low pH. Hence, Sabouraud's medium (initial pH 5.7) is used for their isolation.

Selective media can also aid in taxonomic identification. For example, since *E. coli* uses lactose and *Salmonella* (otherwise very similar metabolically) does not, one way to distinguish them would be to try growth on minimal medium with lactose as sole carbon source. However, it is customary to use an indicator plate with lactose as sole added sugar, plus a rich mixture like peptone (amino acids and peptides). Cells not using lactose still grow on the peptone, but the ones using lactose will give a different pattern of metabolic products, which can be revealed by an oxidation-reduction or acid-base indicator in the agar. Since these tests are often based on the formation of acidic products of sugar metabolism, they are sometimes also loosely called fermentation tests.

Bacterial genetics depends on the detection of rare cells (mutants and recombinants) in cultures otherwise pure or in cultures containing only two different parental types. As in diagnostic or taxonomic bacteriology, media are chosen which distinguish the cells of interest as colonies on appropriate selective plates. The simplest example would be a plate containing an antibiotic, such as streptomycin, for selecting a mutant that is resistant to the drug. For a sensitive organism, if streptomycin is included in a plate otherwise adequate for growth, no colonies would appear other than the rare mutants that are resistant (in this case, about 1 in 10^9 cells).

Growth in Liquid Media

Although primary isolations are done on plates, bacteria are usually grown in a liquid medium. In medical bacteriology and in bacterial genetics, growth media are often rich broths, such as soups from boiled yeast or beef hearts, which contain the ingredients of cells: amino acids and peptides, nucleic-acid degradation products, vitamins and co-factors, and sugars that can be taken up and used. Such undefined mixtures are cheaper than media of defined composition and they usually support faster growth than simpler defined media. Liquid cultures are often agitated mechanically to aerate the cells and to prevent aggregation or settling. Growth may be followed in several ways: (1) viable count; (2) measurement of apparent absorbance or turbidity in a photometer; (3) cell counts in a light microscope; and (4) assay of a particular cell constituent (DNA or protein). For most of these parameters, plots versus time of incubation will look similar: a lag phase, followed by exponential growth, and terminating in a stationary phase (Fig. 4.1).

THE LAG PHASE

The lag phase will depend on the previous history of the inoculum. For example, if the inoculum was old, most of the cells may be nutritionally depleted and some may be dead; the lag will therefore be long. Even if all cells in the inoculum are viable and not starved, they may have come from a medium with a different composition. Cells adapted to that medium might lack certain enzymes that are necessary for growth in the new environment, so new proteins will

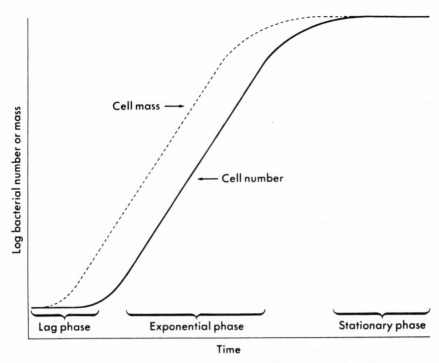

Cell mass →

← Cell number

Log bacterial number or mass

Lag phase   Exponential phase   Stationary phase

Time

Fig. 4.1 Phases of bacterial growth, starting with an inoculum of stationary phase cells. Note that the classic phases, defined in terms of cell number, do not precisely coincide with the phases of changing growth in terms of protoplasmic mass. (From B. D. Davis, R. Dulbecco, H. N. Eisen, and H. S. Ginsberg. 1980. *Microbiology.* 3rd ed. New York: Harper and Row, p. 65.)

have to be made before growth can occur. Or there might be no need for new enzymes, but with the inoculum coming from the stationary phase, the cells may be smaller. Hence there is a longer apparent lag when measured by viable count than measured by mass (Fig. 4.1). Accordingly, one would expect that if cells are taken from the exponential phase and diluted into fresh identical medium, there should be no lag in growth; that is the case.

EXPONENTIAL GROWTH

During exponential growth the rate of increase in cells is proportional to the mass or number of cells already present. For example, for mass M:

$dM/dt = kM.$

Integration gives:

$\ln(M_2/M_1) = k(t_2 - t_1).$

Plotting M on a logarithmic scale versus time on a linear scale will give a straight line with a slope of k. For cells in the exponential phase of growth (also called logarithmic growth), the k will have the same value for whatever parameter is measured (mass/ml, DNA/ml, and so on): when the culture doubles, everything in it doubles. From the second equation, the relationship between doubling time in hours and k is:

$\ln 2 = k \times$ (double time), or $k = 0.69/$(doubling time).

So, if a culture doubles in one hour, k is 0.69 hr^{-1}; if it doubles in 42 minutes, $k = 1$. The constant k is called the first-order growth-rate constant. For a particular species of bacteria, k will depend on the culture medium, oxygen tension, and ambient temperature. Thus, *E. coli* may grow with a doubling time of about one hour at 37°C in a simple medium with glucose as a sole carbon source, but it will grow faster in a richer medium.

Exponential growth in a constant environment is the ideal situation for doing many experiments. The conditions cannot be readily fulfilled in a batch culture, because as the cell density reaches a significant level, nutrient depletion and waste accumulation may become appreciable. As the environment changes, so will the composition of the cells. To maintain a steady growth rate in a constant environment, cells can be cultivated in an apparatus called the **chemostat** into which fresh medium containing a limiting nutrient is continuously pumped; the volume of the culture is prevented from expanding by an overflow channel. The continuous dilution allows the cell density to remain steady and growth to occur in a constant environment.

Fortunately, for most measurements in bacteriology, it is not essential that cultures be in logarithmic phase or steady-state growth. Thus, if one is interested in whether a culture contains a particular

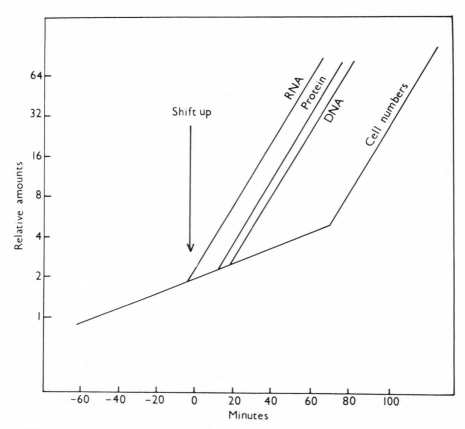

Fig. 4.2 Schematic representation of a shift-up. A culture of *E. coli* growing at 37°C in a synthetic medium at a rate of one doubling per hour is shifted at time zero to nutrient broth, where the rate is three doublings per hour. Since growth is exponential, the semilogarithmic plot of the data includes in a single line the *relative* increment of the various components before the shift. After the shift the synthesis of RNA, protein, and DNA changes to the new, faster rate in that order. Subsequently the rate of cell division increases to the same new value. The size and composition characteristic of the new medium is reached only after 70 minutes from the time of the shift. (From J. Mandelstam and K. McQuillen. 1973. *Biochemistry of Bacterial Growth.* 2nd ed. New York: Wiley, p. 148.)

enzyme, usually the enzyme will still be in the cell after growth has stopped.

Certain bacterial characteristics vary considerably as a function of the growth rate of a culture. The most notable ones are the average size of the cells and the relative quantity of ribosomes (Table 2.1). Fast-growing cells not only are on the average larger than slow-growing ones, but are also richer in ribosomes and may contain two or more copies of the same chromosome. Thus, when cells are shifted from one medium to another that allows a different growth rate, the shift is followed by an adjustment in the rates of macromolecular synthesis until the appropriate composition is attained (for example, see Fig. 4.2). The general rule that fast-growing cells have more ribosomes than slow-growing cells is true for higher cells, too.

SYNCHRONOUS GROWTH

The growth of bacterial cultures reflects the sum of the growth of individual cells. In a steady-state culture many growth characteristics of the individual cells can be inferred from properties measured on the whole population. But in such cultures it is difficult to learn

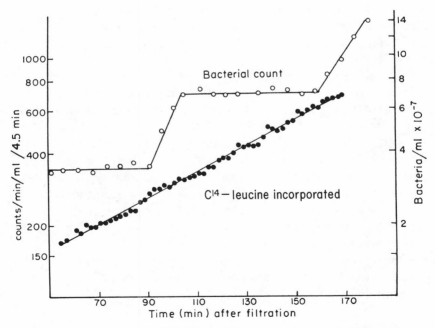

Fig. 4.3 While the bacterial divisions are synchronized so that the counts go up stepwise, the incorporation of amino acid into protein proceeds linearly. (From F. E. Abbo and A. Pardee. 1960. *Biochimica et Biophysica Acta* 39:478–485.)

about the individual order of events in the growth and division of single cells—for example, how a newly formed rod-shaped cell progresses in an orderly fashion to double its size (by lengthening) and then to divide. For in a batch culture individual cells are in different phases of the growth cycle: some have just divided, some are half-way through, and some are just about to divide. Such a culture is called *asynchronous*. It is possible, but difficult, to arrange synchronous growth in cultures. One could mechanically separate all newly divided cells and follow their growth. For a few divisions synchrony of the whole culture would be maintained (Fig. 4.3), so that the timing of particular events in the cycle would be displayed by the entire population.

STATIONARY PHASE

Growth of batch cultures must eventually stop, either because some nutrient is exhausted, or a toxic product (often acid) reaches an intolerable concentration, or both. In practice, with the media and conditions commonly used, the limitation occurs at about 10^9 cells/ml, that is, when the total volume of cells is still only about 1/1,000 of the culture volume.

What actually happens to cells in the stationary phase has not been a favorite subject for study by bacterial physiologists, since an advantage of using bacteria to study cellular processes is their ability to grow. Nonetheless, in nature as a whole most bacteria cannot be growing. Obviously, cells in stationary phase must either remain viable or die; their fate depends both on the bacteria and on the growth medium. Cultures of *E. coli* are relatively hardy; in stationary phase, viability or cell mass/ml remains fairly constant for hours or days. (But even then there would be continuous slow macromolecular synthesis and turnover.) Many bacteria are less hardy, and viable counts may fall quickly in stationary phase, in some cases because of autolysis (as in *Streptococcus pneumoniae*). Other species of bacteria go into chemical hibernation by forming spores as a response to nutrient depletion.

Sporulation

When the supply of carbon, nitrogen, or phosphorus is limited, certain gram-positive rods (that is, aerobic bacilli and anaerobic clostridia) develop highly resistant, dehydrated forms called endospores or spores, which have no metabolic activity. Bacterial spores are particularly adapted for prolonged survival even under adverse conditions such as heat, drying, freezing, the presence of toxic chemicals, and radiation. The longevity and general resistance of spores are probably attributable largely to their low water content. (Most bacteria that are *not* spore formers can be preserved indefinitely by removing their water in the frozen state and keeping them in a vacuum. This is called lyophilization.)

Many metabolites change concentration under the conditions of deprivation that trigger the commitment to sporulate. Some purine nucleotides (3′,5′-cAMP; 3′,5′-adenosine bistriphosphate) and intracellular proteases seem to be involved in the regulation of sporulation. In particular, the specificity of transcription appears to be shifted for a large number of operons by a replacement or modification of the normal initiation factor of RNA polymerase, σ (see Chapter 5). Antibiotics are produced by microorganisms that sporulate, and there is some evidence that these compounds have a direct role in sporulation.

FORMATION AND STRUCTURE

Spores are unusually dehydrated, highly refractile, and do not take ordinary stains (Gram's, methylene blue). In the light microscope the first visible stage in sporulation is the formation of an area of refractility that gradually increases. The mature spore is completed in 6-8 hours and then is freed by autolysis of the wall of the surrounding cell (the **sporangium**). In a rapidly sporulating culture most cells form spores. Characteristic shapes may be seen in sporulating cells and are an aid in identification (Fig. 4.4). Thus bacilli form a central

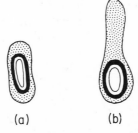

(a)　　　　(b)

Fig. 4.4 Characteristic shapes of spore-containing bacteria useful for identification purposes. (a) *Bacillus*. (b) *Clostridium*.

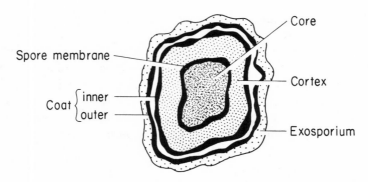

Fig. 4.5 Schematic spore structure.

spore without bulging while some clostridia form a bulging terminal spore.

Sporulation is a process of specialized cell division and morphogenesis. A nucleus and its surrounding cytoplasm become walled off with a series of new protective layers. (The mechanism for eliminating water from the maturing spore is not understood.) The final spore structure is therefore quite different from the original cell (Fig. 4.5). The **core** contains the cytoplasm, the genome, and a number of unusual compounds including large amounts of Ca^{++}-dipicolinate (Fig. 4.6), which seem necessary (but not sufficient) for the heat resistance of spores. The core is surrounded by the **spore membrane,** originating in the former cytoplasmic membrane. The next layer is the **cortex,** which in known cases is a peptidoglycan of a looser structure than that found in the normal cell. Outside the cortex is the coat, which largely consists of proteins rich in cysteine, many of which cross-link different chains by disulfide bridges (like keratins). The coat may be the component responsible for resistance to chemicals, since this property is lacking until the stage of coat formation, and it is permanently absent in mutants with a defective coat. The coat may also be important for the increased resistance of spores to killing by ultraviolet or ionizing radiation. Finally, in some cases there is an outer lipoprotein layer called the **exosporium.**

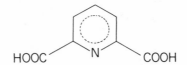

Fig. 4.6 Dipicolinic acid (pyridine-2,6-dicarboxylic acid).

COMPOSITION OF THE CORE

The spore core contains much less material than the vegetative cell, but it must contain everything necessary for resuming growth (see below). It lacks the components of the vegetative cell that are particularly unstable or are readily replenished early in germination, but contains the DNA of one chromosome and small amounts of all the stable components of the protein-synthesizing machinery (though there is no detectable messenger RNA). Amino acids (and their biosynthetic enzymes) are virtually absent; they are supplied early in germination by hydrolysis of a low molecular weight storage protein, which constitutes about 20% of the total protein. There is a pool of ribonucleoside mono- and diphosphates (for building mRNA in germination). Energy for initiating germination is stored as the stable 3-phosphoglycerate, which is readily converted via P-enolpyruvate to ATP.

The striking stability of enzymes in spores depends largely on the environment (dehydration, ionic conditions, and binding of stabilizing factors), since many enzymes purified from spores are identical with those in the parental cells and are not particularly thermostable in solution (though a few enzymes are cleaved by a protease to yield a smaller, more stable enzyme in spores).

Germination

The process of converting a spore into a vegetative cell is often called germination. Though it is much faster than sporulation, three stages can be distinguished: activation, initiation, and outgrowth.

Activation Though some bacterial spores may germinate spontaneously in a favorable medium, others (especially if freshly formed) require activation by some agent, such as heat, low pH, or a sulfhydryl compound. Aging is the most important natural cause. Activation presumably damages the impermeable coat, since grinding with glass powder is also effective.

Initiation This step requires water and a triggering germination agent. Various species respond to various metabolites (for example, alanine, dipicolinate) or inorganic ions (Mn^{++}) which penetrate the damaged coat. Cortical peptidoglycan is hydrolyzed quickly; the cell rapidly takes up water, K^+, and Mg^{++}; and refractility is lost (the usual test for germination) along with resistance to heat and to staining. Most of the energy stored in 3-phosphoglycerate is converted to ATP, and the storage protein is hydrolyzed within minutes, liberating amino acids and more energy.

One way of exterminating spores is by two successive periods of boiling. After the first exposure, the spore will germinate; the second boiling kills the germinated cells.

Outgrowth During the gradual resumption of vegetative growth (outgrowth), protein synthesis increases in rate, as its initial scant machinery is expanded by new protein and RNA. The spore wall becomes a thicker vegetative cell wall. The specificity of RNA polymerase is restored early, whereas DNA synthesis begins later. Eventually, the cell increases in volume and bursts out of the spore coat.

Questions

4.1. The smallest colony visible without magnification is around 0.2 mm in diameter. Assume it is a hemisphere containing cylindrical cells 1 μ in length and 0.5 μ in diameter (1 μ = 0.001 mm). (a) How many cells would such a colony maximally contain? (Recall that the volume of a sphere is $\frac{4}{3}\pi r^3$ and the volume of a cylinder is $\pi r^2 h$.) (b) Using the answer from (a) and assuming a 10-hour period of incubation, what would be the minimal cell doubling time (the time for n cells to grow into 2 × n cells)?

4.2. How might one inoculate a single cell into liquid medium?

4.3. A bacterial culture was diluted in steps of 1 : 10 in a buffer, and 0.1 ml portions were spread on plates. After incubation, the number of colonies was: 8 from the 10^{-8} dilution, 113 from the 10^{-7} dilution, and 650 from the 10^{-6} dilution. What was the viable count in the original culture?

4.4. Suppose a culture were inoculated at 10^4 cells/ml in a medium that allowed a growth rate of one doubling/hour. What would be the cell concentration in three days, if there were no physical restrictions?

4.5. In a starved culture only 10% of the cells are viable. A sample of this culture is then inoculated into a rich medium. Supposing that all the survivors grow exponentially, draw the growth curve as measured by turbidity and as measured by viable count.

4.6. Bioassay means the use of a biological system to test for the presence of a particular compound. Thus, to assay histidine with S. *faecalis* one makes a medium containing a carbon source (such as glucose) and all the required growth factors with the exception of histidine. This medium would be added to a series of tubes, some with known amounts of histidine, and others with portions of the unknown sample. S. *faecalis* is inoculated to each tube and allowed to grow to stationary phase. Suppose that in this system, 10 μg histidine/ml gives maximum growth. Draw the growth curves for 1, 5, 10, and 20 μg histidines/ml. Also plot the final growth yield versus the concentration of histidine in the medium. The advantage of this type of assay is that it would work with a sample such as blood which is full of other organic material. It is both sensitive and specific. Most chemical assays would require prior purification.

4.7. The results of two different growth experiments with the same medium are described in Figure Q4.7. Which are *reasonable* explanations for the difference in stationary phase titer?

(a) Organism 1 produces an acidic end-product of metabolism which organism 2 does not.

(b) Organism 1 has a faster doubling time than organism 2. Therefore organism 1 uses up nutrients sooner, and growth stops at a lower cell density.

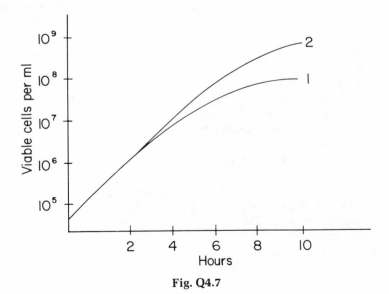

Fig. Q4.7

(c)   Organism 1 is much larger in size than organism 2.

(d)   Organism 2 can catabolize a broader spectrum of compounds than organism 1.

4.8. "Kligler's iron agar" is a typical diagnostic medium used for distinguishing certain types of enteric organisms. It contains amino acids, 0.1% glucose, 1% lactose, phenol red (an acid-base indicator which turns yellow below pH 5), ferric citrate, and Na^+-thiosulfate (thiosulfate cleavage to H_2S will give a Fe_2S_3 precipitate which is black). The medium comes as slants in tubes and is inoculated with the isolate by streaking on the flat surface (for aerobic growth) and stabbing with a needle into the agar (for anaerobic growth). A typical result after incubation overnight would be:

Bacteria	Surface	Butt	Precipitate
E. coli	yellow	yellow	no
Salmonella	red	yellow	yes

Interpret the findings.

Further Reading

Dykhuizen, D. E., and D. L. Hartl. 1983. Selection in chemostats. *Microbiological Reviews* 47:150–168.

Herbert, D., R. Elsworth, and R. C. Tellings. 1956. The continuous culture of bacteria: a theoretical and experimental study. *Journal of General Microbiology* 14:601–622.

Ingraham, J. L., O. Maaløe, and F. C. Neidhardt. 1983. *Growth of the Bacterial Cell.* Sunderland, Mass.: Sinauer.

Levinson, H. S., A. L. Sonenshein, and D. J. Tipper, eds. 1981. *Sporulation and Germination.* Washington, D.C.: American Society for Microbiology.

Maaløe, O., and N. O. Kjeldgaard. 1966. *Control of Macromolecular Synthesis: A Study of DNA, RNA and Protein Synthesis in Bacteria.* New York: W. A. Benjamin.

Mandelstam, J., and K. McQuillen, eds. 1968. *Biochemistry of Bacterial Growth.* New York: Wiley.

Mendelson, N. H. 1982. Bacterial growth and division: structures, forces, and clocks. *Microbiological Reviews* 46:341–375.

Monod, J. 1949. The growth of bacterial cultures. *Annual Review of Microbiology* 3:371–394.

Young, M., and J. Mandelstam. 1979. Early events during bacterial endospore formation. *Advances in Microbial Physiology* 20:103–162.

5

Gene Expression and Regulatory Mechanisms

THE *E. coli* chromosome comprises about 3000 to 4000 different genes, each of which encodes information for the synthesis of a protein or RNA molecule. However, for any given cell, no more than half of these genes are significantly expressed. A variety of factors together determine which genes will be expressed, and one of the most important of these factors is the chemical composition of the growth medium. For example, if a culture of a bacterium with simple nutritional requirements, growing exponentially on glycerol as the sole source of carbon and energy, is shifted to a medium in which glycerol is replaced by lactose, there will be a lag period of about half an hour before growth resumes. The same is true if the cells are switched from a lactose to a glycerol medium. This is because cells growing on glycerol do not have the special catabolic ability to handle lactose, and vice versa; a certain amount of time is required to synthesize the necessary enzymes in the catabolic pathway.

Moreover, when a culture of a bacterium with simple nutritional requirements is fully grown in a minimal medium with glycerol or glucose as carbon and energy source, some fermentation products may be found in the medium (particularly if aeration was limiting), but it will contain little or none of the major biosynthetic products of small molecular weight, such as amino acids or nucleic acid bases. This is because the synthesis of these compounds is precisely regulated by the cell to give just the amount needed in macromolecular synthesis.

And finally, if the same organism is grown in a rich medium in which amino acids or nucleic acid bases are provided, growth will be faster than in minimal medium, because the cells use many of these exogenous compounds directly (when there are cellular transport mechanisms), instead of synthesizing them de novo.

These examples hint at elaborate control mechanisms by which bacterial cells regulate their catabolic and biosynthetic pathways so as to avoid wasteful over-production. Regulation often takes place at the level of transcription—where some genes are activated while others are not. But regulation can also take place at the level of translation, and at the level of protein function after proteins are made.

General Features of Transcription and Translation

TRANSCRIPTION

The first step in gene expression—**transcription**—always involves forming a molecule of RNA that corresponds to the base sequence in the segment of DNA being transcribed. This growing strand of RNA forms according to the same rules of base pairing that operate in DNA replication (A-T, T-A, G-C, C-G), except that RNA has no thymine; uracil, another pyrimidine, is found instead. Either strand of the DNA may serve as the template for transcription (that is, act as the **sense strand**), depending on the particular gene. The template DNA strand is always read in the 3' to 5' direction, and the RNA is always elaborated in the 5' to 3' direction. It therefore follows that with respect to the double-stranded circular chromosome, some genes are copied in the clockwise direction and some in the counterclockwise direction.

In the case of the structural genes for tRNA (t for transfer) and rRNA (r for ribosome), the transcripts themselves are the products, although further modification of these RNAs (such as methylation of certain bases) has to first take place before they become units in the protein-synthesis machinery. In the case of mRNA (m for messenger), on the other hand, the transcript serves merely as an intermediary for further information processing, namely the **translation** of each group of three bases into one of the twenty amino acids that make up proteins (see Fig. 5.1). Since prokaryotic genes encoding proteins are not interrupted by regions not specifying amino acid sequences, in contrast to the vast majority of eukaryotic genes (cod-

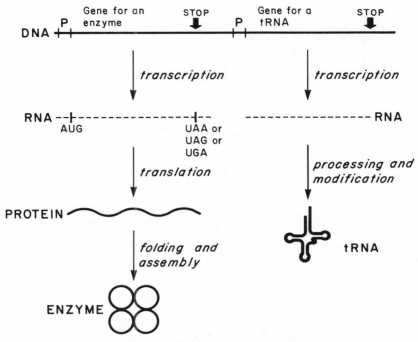

Fig. 5.1 Schematic representation of the expression of two kinds of structural genes. On the left, the gene codes for an mRNA which is then used as a template for protein synthesis. The amino acid sequence of the polypeptide contains the information for correct folding and tetramer assembly. On the right, the gene codes for a tRNA. Translation is unnecessary, but maturation of a functional molecule requires chemical changes. Internal complementarity of base sequences causes stem-loop formation.

ing regions being referred to as **exons** and noncoding regions referred to as **introns**), the mRNAs can be translated directly without splicing.

The enzyme that plays the central role in the transcription process is a DNA-dependent **RNA polymerase,** also called transcriptase. The "core" of RNA polymerase is composed of two α subunits, one β subunit, and one β' subunit. The core enzyme can combine with an additional subunit called σ to become the **holoenzyme** (the complete enzyme). This last subunit helps the polymerase complex to locate the region of DNA where transcription is to be initiated. This critical region of DNA, which strongly interacts with the holoenzyme (probably causing local separation of the two sister DNA strands), is called the **promoter** (Fig. 5.2). Most promoters consist of a 7-base-pair sequence rich in AT. Once the promoter and the enzyme are posi-

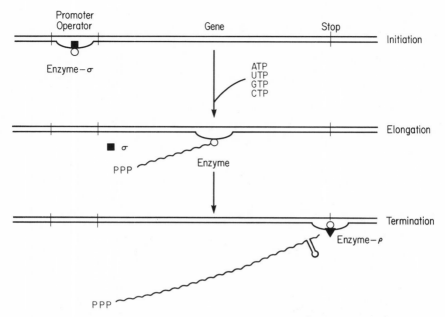

Fig. 5-2 Schematic representation of transcription. Three forms of the RNA polymerase complex participate respectively in mRNA chain initiation, elongation, and termination. The initiation requires the recognition of the promoter by the enzyme (O) with the aid of the σ subunit (■), which dissociates from the enzyme complex once chain elongation begins. Chain elongation proceeds with the incorporation of ribonucleotides at the expense of ribonucleoside triphosphates. For certain genes, like this one, termination involves the protein ρ (▼) which interacts with the polymerase and causes dissociation of the enzyme from the transcript and template. Inverted sequences of complementary bases at the transcript terminus allow the mRNA to fold back upon itself, forming a stem loop which is probably a part of the termination signal.

tioned in register, transcription begins and the σ subunit falls off, leaving the core enzyme to carry out elongation of the RNA chain. Transcription is terminated by the falling off of the core enzyme, and this is signaled by a special base sequence *ter* (for termination). The terminal segment of the mRNA typically contains base sequences that permit hairpin (also referred to as stem-loop) formation by internal base pairing. Possibly this structure serves as a signal for the RNA polymerase to dissociate from the DNA template. However, some terminations are dependent upon a special protein (see below).

In many cases, the decision whether a gene is to be transcribed or terminated is not made only by RNA polymerase and particular base sequences of the DNA template. For certain genes, transcription can occur only when a general and/or specific **activator protein** is present to initiate the event or when there is no obstruction by a specific protein (**repressor**). Likewise, for some genes, transcription termination can be assured only with the participation of an auxiliary protein called ρ or rho (Fig. 5.2). In special cases, the absence of ρ at the critical moment and at a specific position of the template DNA leads to continued transcription of the next gene(s) downstream.

The concentration of an mRNA species in a cell reflects the balance between the frequency of transcription and the rate of degradation. The half-life of an average bacterial mRNA at 37°C is about 2 minutes. In general, degradation of mRNA occurs enzymatically from the 3′ end. Continuous translation of an RNA molecule prolongs its existence because of physical protection by the ribosomes. Special base sequences in the mRNA may also be important. For instance, a stem-loop structure at the 3′ end of certain mRNA species was shown to afford protection against 3′ RNA nucleases.

TRANSLATION

The key agent that translates the code in the mRNA into a proper order of amino acids in a polypeptide is the tRNA. A group of enzymes, the aminoacyl-tRNA synthetases, are responsible for charging the tRNAs with their cognate amino acids by covalent linkage of the carboxyl group of the amino acid to the 3′ terminus of the tRNA. The loaded tRNA is matched with the base triplet (**codon**) in the mRNA by complementarity. For this reason the tRNA was once called also the adapter RNA. The ribosome provides the framework on which accurate reading and the successive addition of amino acids take place.

Since the bacterial ribosome has a sedimentation constant of 70S (molecular weight of about 3 million), it is often referred to as the 70S ribosome. This large particle has a diameter of about 18 nm (10^{-6} mm) and contains ⅔ rRNA and ⅓ protein. The ribosome reversibly dissociates into a 50S subunit and a 30S subunit at one point of the translational cycle. During the course of protein synthesis — polypeptide chain initiation, elongation, and termination — the 70S ribosome holds on to the mRNA until the completed protein is made.

Then the ribosome dissociates into its two subunits (one 50S and the other 30S) before a new cycle of translation begins.

Initiation The 30S ribosomal subunit, along with a special tRNA bearing an N-formylmethionyl group (fMet-tRNA$_f^{met}$), combines at a specific site with the mRNA to form an **initiation complex** (Fig. 5.3). This domain of the mRNA includes a purine-rich stretch typically represented by AGGAGGU (the Shine and Dalgarno sequence) situated, on the average, 7 bases away from the 5′ end of the initiation codon AUG (or less often, GUG). The sequence is recognized by its complementarity to the 3′ end of the 16S rRNA of the 30S ribosomal subunit. Then a 50S ribosomal subunit attaches to the complex to form the 70S ribosome, which mediates polymerization. Three accessory proteins (termed initiation factors IF1, IF2, and IF3) also take part in the reaction.

Elongation The addition of each amino acid during the polymerization process involves the same sequence of events: (1) *Recognition.*

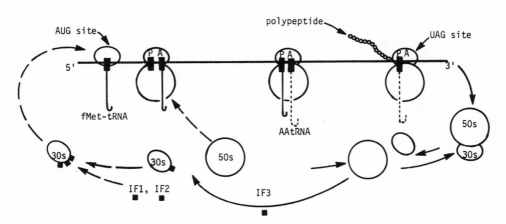

Fig. 5.3 Schematic representation of translation (protein synthesis): polypeptide chain initiation, elongation, and termination (see text).

An appropriate molecule of aminoacyl-tRNA binds noncovalently to the A (or aminoacyl) site of the ribosome according to the codon of the mRNA. The amino acid on this tRNA will be added to the growing peptide, which is attached to another molecule of tRNA at the P (or peptidyl) site of the ribosome. The recognition step requires the participation of two more accessory proteins (termed elongation factor EFTu and EFTs) and hydrolysis of GTP. (2) *Peptide transfer.* The peptide at the P site is then transferred to the amino acid attached to the tRNA molecule at the A site, forming a new peptide bond and thereby lengthening the peptide chain by one aminoacyl residue, and releasing the tRNA from the P site. Peptide transfer is catalyzed by the 50S ribosomal subunit itself; no accessory proteins are required. (3) *Translocation.* Following peptide transfer, the peptide-bearing tRNA moves to the P site and the mRNA moves with it. Translocation requires one accessory protein (termed elongation factor EFG).

Termination By repetition of the recognition, peptide transfer, and translocation steps, successive aminoacyl residues are added to the peptide chain in the order specified by the sequence of codons in the mRNA molecule. The process continues until a codon (UAG, UAA, or UGA) that causes the release of the completed peptide is reached. The termination process requires the intervention of a protein called a release factor (RF) which hydrolyzes the bond between the polypeptide and tRNA. The ribosome is then released from the mRNA by another protein, the ribosome release factor (RRF). Dissociation of the ribosome into its subunits is promoted by binding of IF3 to the 30S subunit; then binding of IF1 and IF2 to that subunit prepares it for the next initiation. Recognition and translocation each require energy, provided by GTP hydrolysis. The energy for peptide transfer is provided by the ester bond between the peptide and the tRNA.

Negative Control of Transcription in Catabolic Pathways

THE LACTOSE SYSTEM

The system governing lactose utilization is one of the best understood genetic regulatory mechanisms in prokaryotes. Many of the general characteristics of prokaryotic gene expression were discovered in the lactose system.

When *E. coli* is growing on lactose as carbon source, two proteins essential for lactose metabolism are abundantly synthesized. One is β-galactoside permease, which is necessary for transport and accumulation of lactose in the cell, and the other is β-galactosidase, an enzyme essential for the hydrolysis of lactose. These proteins are encoded by two neighboring genes: *lacZ* for β-galactosidase and *lacY* for the permease. There is a third gene, *lacA*, which encodes a transacetylase whose function appears to be the conversion of a nonmetabolizable β-galactoside for excretion. Transcription of these genes—in the order of ZYA—is initiated when RNA polymerase recognizes and binds to a site near *lacZ*, called *lacP*, the promotor. The *lacP* region was defined by mutations near, but not within, the *lacZ* gene that affect cellular levels of β-galactosidase, β-galactoside permease, and transacetylase.

During transcription the genes *lacZ, Y,* and *A* are copied onto a single mRNA molecule. A unit of DNA that is transcribed onto a single mRNA molecule is called an **operon.** An operon can code for one or more polypeptides. Genes of a given metabolic pathway are often clustered in a single operon in bacteria, but so far no example of a multigenic operon has been found in eukaryotes, from yeast to humans.

If lactose is absent from the medium, transcription of the *lac* genes occurs rarely, as though by accident. Since the average mRNA half-life in bacteria is only a few minutes, such rare transcription leads to very limited synthesis of the proteins in the lactose pathway. However, such seemingly accidental synthesis accounts for the presence of *basal* levels of the lactose proteins. (Bacteria maintain basal levels of all inducible catabolic enzymes.) These low levels of enzymes allow a few molecules of lactose to be transported into the cell when lactose is first introduced into the medium. These molecules then induce the increase of enzyme production in the lactose pathway, in the following sequence of events.

Enzyme induction involves influencing a protein called the *lac* **repressor,** which is specified by the gene *lacR*. (R is used in this work to represent repressor or regulator. Conventionally, *lacI* is used for historical reasons, I standing for induction.) The repressor itself is produced at a low and unregulated rate. In the absence of lactose, the repressor binds to a DNA site located between the promotor, *lacP*, and *lacZ*, thus preventing the initiation of transcription. The site to which the repressor binds is called the **operator,** or $lacO$ (Fig. 5.4). However, when lactose is brought in, β-galactosidase converts it by a side reaction to an isomer, allolactose, which interacts with the repressor protein. The conformation of the $lacR$ protein is changed so that it no longer binds at the operator. The use of allolactose rather than lactose itself as the actual **inducer** probably provides the cell with an additional screening mechanism to ascertain the identity of a metabolizable sugar. (The first screening by β-galactoside permease is not entirely reliable because of the limited specificity of this membrane protein.) False induction is not only wasteful of protein synthesis, but might result in the active accumulation of a nonmetabolizable analog. There are nonmetabolizable analogs of lactose that have inducing properties (such as isopropylthio-β-D-galactoside [IPTG]) and also substrates of the enzyme that are poor inducers.

Induction of expression of the lac operon is rapid. High rates of synthesis of the encoded proteins begin within a few minutes of addition of lactose to a culture. When lactose is removed, new protein synthesis from these genes stops, again within a few minutes. The lactose proteins already present are diluted out by further growth (Fig. 5.5). (Protein turnover does occur in bacteria but at low rates.)

The lactose system thus provides an example of **negative control** of gene expression by repressor. The gene $lacR$, which encodes the

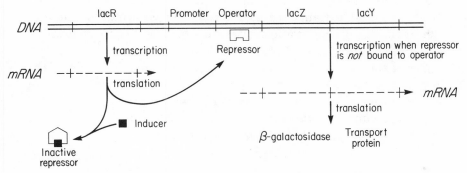

Fig. 5.4 The lactose operon of E. coli and the regulation of its expression. The $lacR$ gene codes for a specific repressor protein that binds to the lac operator site (sometimes referred to as $lacO$). Binding of the repressor by the inducer stabilizes the inactive conformation of the regulatory protein and frees the operator to allow transcription. A third gene ($lacA$) distal to $lacY$ is not shown in the diagram, because the protein is not necessary for the utilization of lactose.

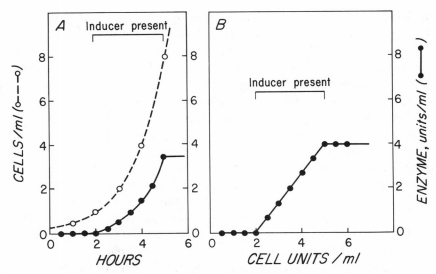

Fig. 5.5 The kinetics of induction of β-galactosidase. Cells of E. coli were growing on glycerol. Open circles represent measurements of growth; solid circles, enzyme units.

repressor, prevents transcription and expression of the genes specifying the metabolic pathway; inducers work by sequestering the repressor in an inactive state. It is possible to select mutants in which the repressor has been altered or eliminated, resulting in the continuous transcription of the *lac* operon with or without an inducer. Such bacterial variants are called **constitutive mutants.** Another class of constitutive mutations affects the operator itself, lowering its ability to bind repressor; these are called O^c mutations.

In yet another type of mutant, the repressor cannot be held by the inducer in the inactivated state. Under these conditions repressor binds to the operator even in the presence of lactose, rendering the *lac* system uninducible or superrepressed. (The phenotype of such mutants is Lac⁻; the mutant gene is called *lacR*ˢ.)

Positive Control of Transcription in Catabolic Pathways

THE ARABINOSE SYSTEM

Repressor control is only one of a variety of mechanisms whereby bacterial gene expression is regulated, even among inducible catabolic pathways. In the case of the L-arabinose pathway in E. coli, structural genes for the degradative enzymes are said to be under **positive control,** because a regulatory gene, araR (conventionally araC), codes for a protein which is necessary for the expression of the genes determining arabinose metabolism (Fig. 5.6). Arabinose holds the regulatory protein in an active conformation allowing binding to the promoter site; this binding stimulates transcription initiation by RNA polymerase. (Interestingly, in the absence of the inducer, the regulator protein behaves as a repressor.) Mutants in which the araR gene is deleted and the protein is not produced cannot metabolize arabinose. In contrast, deletion of the lacR gene results in continuous production of the lactose-catabolizing enzymes even in the absence of the inducer. A number of inducible catabolic pathways have been examined for their regulatory mechanism; they appear to be evenly divided between positive and negative control.

Another interesting comparison of the lacR and araR genes revolves around the question of whether or not the synthesis of regulatory proteins itself is regulated. In contrast to the lacR gene, which seems to be expressed at a low constitutive level, the araR gene seems to be repressed by its own product (**autogenous regulation**). By

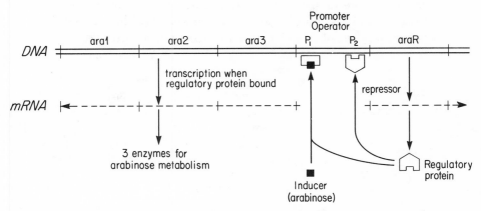

Fig. 5.6 The arabinose operon of E. coli and the regulation of its expression. The araR gene codes for a specific regulator protein which can be in two forms. The inducer shifts the protein into the one that activates the promoter of the structural genes for arabinose utilization. When the protein is in the other form, it actually acts as a repressor of the structural operon (detail not shown). In either form, the protein represses the expression of its own gene (autogenous regulation). Transcriptions of the two operons are in opposite directions, in contrast to the lac system.

both controlling the arabinose operon and repressing its own synthesis, the $araR$ protein also illustrates the ability of regulatory proteins to serve more than one function. The lac repressor, by contrast, has only one function, the repression of the $lacZYA$ operon.

Control of Transcription in Biosynthetic Pathways

In the inducible catabolic pathways discussed, the enzymes necessary for degradation are produced in substantial amounts only when the particular carbon source is available. With biosynthetic pathways, an opposite situation exists: if an amino acid that can be used is available in the medium, the level of the enzymes necessary to produce that product endogenously is diminished. The regulatory system that prevents wasteful assembly of the enzymes in a biosynthetic pathway when there is an external supply of the product is called **end-product repression** (Fig. 5.7).

THE ARGININE PATHWAY

Enzymes for the biosynthesis of arginine are at their highest levels when the intracellular concentration of arginine is low. But when a repressor protein is complexed with arginine, it specifically binds to operators of genes of this biosynthetic system, preventing transcrip-

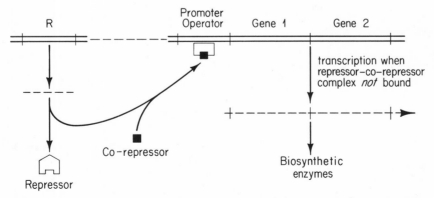

Fig. 5.7 The regulation of expression of genes specifying a biosynthetic pathway. An important variation in this scheme is that the small molecule co-repressor serves to stabilize the regulatory protein in the repressive form.

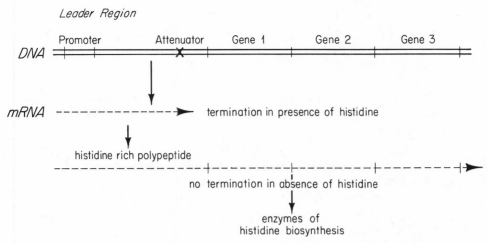

Fig. 5.8 The regulation of expression of the histidine operon encoding enzymes for histidine biosynthesis by attenuation or abortive transcription. Transcription of this operon is not regulated by activator or repressor proteins but by whether or not the process continues into the region containing the structural genes. Abundance of histidine reflected by the high proportion of the cognate tRNA charged by the amino acid favors premature transcript termination. The leader contains a region that codes for a polypeptide rich in histidine but without any catalytic function. The way in which the rate of synthesis of this polypeptide influences attenuation is shown in Figure 5.9.

tion. Thus arginine may be considered a **co-repressor** in this case. Inactivation of the repressor by mutations in the *argR* gene results in the constitutive synthesis of the enzymes in the arginine pathway.

THE HISTIDINE PATHWAY

The synthesis of the enzymes in this pathway responds in the same way to histidine as the synthesis of the enzymes of the arginine pathway responds to arginine. But despite the superficial similarity, the mechanism of control of histidine biosynthesis is totally different. There are 9 genes for histidine biosynthesis, all clustered in a single operon (Fig. 5.8). Whether histidine is abundant or not, RNA polymerase initiates transcription of the operon at regular time intervals. If the histidine supply is deficient, the mRNA synthesis continues until all of the genes are transcribed. On the other hand, if the histidine supply is adequate, RNA synthesis is terminated abortively. The product is a **leader transcript** of about 200 nucleotides in length which has two unusual properties: (1) it contains a set of base sequences with dyad symmetries (A to F) that can pair in different ways to form alternative stem-loop structures; and (2) it encodes a leader peptide of 16 amino acids which include seven histidines in tandem (Fig. 5.9). The state of histidine supply is signaled to the transcriptional process in the following way:

When the supply of histidyl-tRNA in the cell is abundant, the first ribosome that translates the leader peptide does not lag far behind the RNA polymerase (Fig. 5.9:I). This prevents the formation of the stem-loop structure AB, but allows two stem loops, CD and EF, to form. The EF structure (together with a series of uracils which follow) is a typical transcription termination signal in prokaryotic systems.

On the other hand, when histidyl-tRNA is deficient, the ribosome is stalled in the region of the *his* codons, thus preventing sequence A from pairing with the complementary sequence B. But it happens that sequences B and C also have a dyad symmetry. The unavailability of A thus allows the formation of the stem loop BC. The preemption of C by B leaves sequence D unpaired until sequence E is made. D and E also happen to have complementary stretches, which permits the formation of the stem loop DE. This leaves F without a partner to form the **attenuator** EF (Fig. 5.9:II). Unhindered by formation of the attenuator stem-loop, transcription then continues into the *his* operon proper, which contains the structural genes encoding the enzymes for histidine biosynthesis.

If for any reason there is an overall deficiency of amino acids, the ribosome becomes stalled before even reaching the region of tandem *his* codons. In this situation, the formation of the attenuator signal is again possible, thus aborting further transcription (Fig. 5.9:III) and economically curtailing the production of histidine biosynthetic enzymes.

In contrast to regulation by repression (negative control) or activation (positive control), **attenuator control** requires no structural gene for a regulatory protein. The trade-off for this saving is the continuous, but modest, consumption of metabolic energy in the production of the leader transcripts and peptides. It appears that in enteric bacteria transcription attenuation is used for regulating most of the amino acid biosynthetic systems. (There is evidence for similar

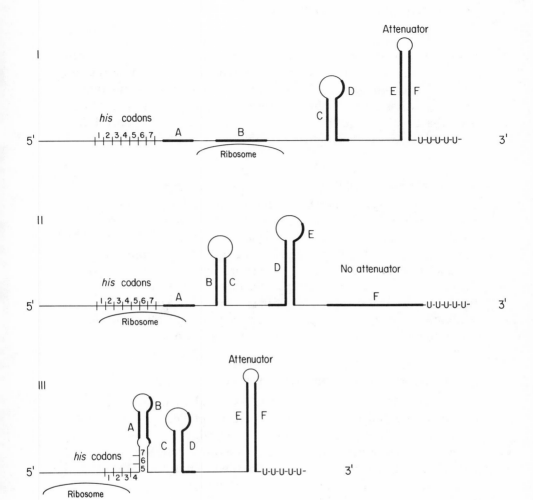

Fig. 5.9 Attenuation of histidine operon expression. The promoter-proximal segment of the *his* operon contains a leader region with base sequences that allow several modes of internal base pairing of the transcript. The 5′ portion of this leader transcript codes for a polypeptide with seven histidines in tandem. The stem-loop EF, or the attenuator, serves as a signal for transcription termination. Whether or not that structure is formed depends on the relative rate of transcription and translation. When all the aminoacyl-tRNAs are sufficient, rapid translation allows the attenuator to form, thus preventing expression of the structural genes for histidine biosynthesis (I). When there is a deficiency of histidyl-tRNA in particular, the ribosome is stalled in mid-translation of the leader peptide. This prevents the formation of EF (II). On the other hand, when there is general deficiency of aminoacyl-tRNAs, the ribosome is stalled before the region rich in histidine codons. EF formation is again possible (III).

regulatory mechanisms in eukaryotic organisms such as yeast.) In principle, this kind of control may be exploited for modulating the expression of any gene whose product can influence the rate of protein synthesis (such as functions related to growth rate), and there is increasing evidence that this is the case. Attenuator control may very well have evolved before repressor or activator control.

THE TRYPTOPHAN PATHWAY

The genes of tryptophan biosynthesis provide an example of an operon controlled by two different pathway-specific regulatory mechanisms. In this system there is a protein repressor specified by the trpR gene, which in combination with tryptophan, prevents transcription of the trp operon. However, there is also an attenuator region at the beginning of the trp operon which can cause transcription termination when high levels of tryptophan are present. The two mechanisms, repressor–operator interaction and attenuation, respond to different degrees of tryptophan starvation.

REPRESSIVE CONTROL OF DIPHTHERIA TOXIN SYNTHESIS

Diphtheria toxin is synthesized and excreted by Corynebacterium diphtheriae when its environment is iron-deficient. A single toxin molecule is lethal within a mammalian cell. In a sense the toxin serves a biosynthetic function by making the iron in the animal cell available to the pathogen. Synthesis of the toxin is under the control of a repressor with Fe^{2+} as the co-repressor.

Regulons

Not all genes specifying a catabolic or biosynthetic pathway are arranged in a single operon. For example, the glp genes, encoding the proteins for the utilization of glycerol, glycerol 3-phosphate, and glycerol phosphodiesters, are distributed in four operons located in different parts of the chromosome. Yet the synthesis of proteins from all four operons is induced by glycerol (the true inducer in this case being glycerol 3-phosphate) and all the operons are subject to the control by the same repressor molecule, the glpR protein. This implies that all four operators share common base sequences for the recognition of the common regulatory protein.

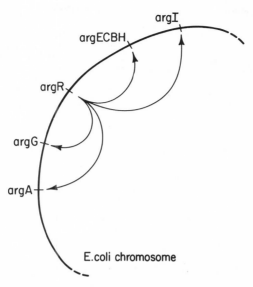

Fig. 5.10 The simultaneous regulation of the expression of several operons of the arginine biosynthetic pathway by a single repressor in *E. coli*. The collection of the operons constitutes a regulon. The gene *argR* codes for a specific repressor. The other genes code for the biosynthetic enzymes.

A similar situation is also found in the arginine biosynthetic pathway, where the structural genes are scattered on the chromosome, most of them as operons containing single genes, all subject to the *argR* repressor protein (Fig. 5.10). A genetic system with several operons under the control of a single regulatory protein is called a **regulon.**

In the case of a multigenic operon, the several proteins coded for by the single transcript vary *coordinately:* if one is tenfold induced, so are the others. In contrast, the genes of a regulon, since they have separate operators, can respond *noncoordinately* to their repressor.

General Regulatory Mechanisms Affecting Transcription

In addition to the pathway-specific controls discussed above, an operon may be subject to a number of more general signals that affect transcription. For example, the inducible histidine degradative operon is subject to both catabolic repression and ammonia repression (see below). Other operons are controlled by cyclic AMP as well as some general signal reflecting whether the medium is aerobic or anaerobic. Many general signals controlling transcription may still be discovered; in the following section we will discuss a few which are already known.

CATABOLITE REPRESSION AND CYCLIC AMP

In bacteria such as *E. coli*, the enzymes in many catabolic pathways are found in lower amounts when glucose is present than under many other growth conditions. Monod showed that if *E. coli* was grown on a mixture of glucose and lactose, lactose was not used until the glucose was exhausted, and only then was β-galactosidase synthesis induced (Fig. 5.11). The presence of glucose in the growth medium has similar effects on other inducible catabolic pathways. There are several mechanisms that explain the effect of glucose.

First, in order for lactose to induce β-galactosidase and be used, it must gain entrance into the cell via the *lac* transport system present at a very low basal level. Glucose, through its own transport system, impedes the function of the *lac* transport system, thus preventing induction. However, if the problem of specific induction is avoided by using a *lacR⁻* (constitutive) mutant, the level of expression of the *lac* system can still be influenced by the composition of the medium: in cells grown on succinate the level of β-galactosidase may be 100; for glycerol, 80; for glucose, 25; and for glucose-6-P, 10. This phenomenon was named **catabolite repression,** because large reductions in β-galactosidase level occur even with succinate as carbon source, provided that biosynthetic pathways are somehow slowed down (for example, by partial starvation for amino acids in appropriate biosynthetic mutants). It seems that the presence of high intracellular concentrations of intermediates feeding into pathways for carbon assimilation and energy generation prevents induction of superfluous catabolic enzymes.

A partial explanation for this effect came with the discovery that cyclic AMP, as well as inducer, is necessary for the expression of operons repressible by catabolites. Subsequent studies showed that cAMP acts in conjunction with a cAMP-binding protein (CAP) to stimulate transcription at the *lac* promoter. In the absence of CAP and cAMP, RNA polymerase interacts only weakly with its binding site on the promoter. But when the CAP-cAMP complex binds to a

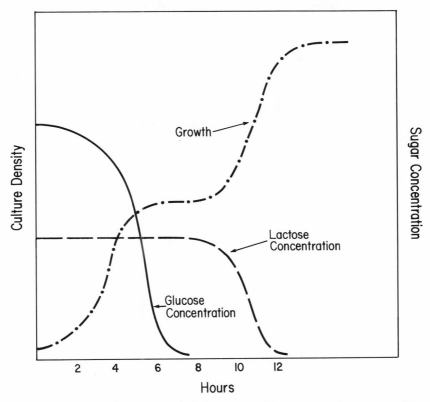

Fig. 5.11 Biphasic (diauxic) growth in a medium containing two carbon sources. Glucose is preferentially utilized; its presence prevents the induction of the lactose enzymes. The enzymes for glucose utilization are generally present at high basal levels in E. coli.

site on the DNA adjacent to the polymerase site (both of which are a part of the promoter), transcription is initiated (Fig. 5.12). This is a type of positive control. Although CAP is found in cells grown on any medium, the level of cAMP in the cell differs considerably between media and is lowest in the situation of greatest catabolite repression. In an unknown way, the metabolism of glucose (or, as mentioned, even of succinate when certain biosynthetic reactions limit the rate of growth) controls the cAMP concentration. That control might involve the level of adenyl cyclase (product of the cya gene), its catalytic activity, or the rate of cAMP excretion into the medium.

The cAMP-CAP control system is classified as a **general (global)**

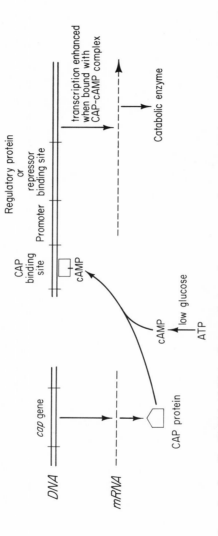

Fig. 5.12 The role of cyclic AMP (cAMP) in the regulation of expression of genes specifying catabolic enzymes. This kind of control is called catabolite repression. The CAP protein in complex with cAMP serves as a general activator for transcription. Though necessary, it is not a sufficient condition for transcription. Full expression of the gene depends also on the specific activator being present or the specific repressor being absent.

control system, because it affects a large number of degradative operons (such as *lac*, *ara*, and g*lp*). Moreover, catabolite repression is limited not just to degradative pathways. In *E. coli* the formation of the flagella can also be repressed by glucose. Thus, a mutation in the *cap* gene (encoding the CAP protein) or *cya* gene can shut off a large number of operons in the bacterial chromosome. Many other kinds of operons, however, are unaffected by catabolite repression. In fact, mutants unable to make cAMP or lacking CAP protein can be selected. Such mutants will grow on a mixture of amino acids and other nutrients.

The DNA base sequence of the *lac* controlling region is now known (Fig. 5.13). Both the operator (*lacO*) and the CAP site have regions of twofold symmetry. The RNA polymerase interaction site contains a region high in AT surrounded by two regions high in GC. Some of these features are proving to be common in control regions of a number of different operons. Mutations have been found that alter specific base pairs in each of the three regions.

Catabolite repression and cAMP control of sugar degradative pathways is common but not universal in bacteria. For instance, cAMP control is absent in *Bacillus subtilis*. Moreover, the hierarchy of preferred sources of carbon and energy depends on the organism. Thus, certain species of Pseudomonas would utilize succinate before glucose when both compounds are presented simultaneously. Again, preferential utilization of one substrate is assured by the failure of induction of the pathway for the other substrate, although the precise mechanisms in those cases are not yet known.

ppGpp

When a mutant unable to synthesize a particular amino acid is starved of that amino acid, protein synthesis stops. In addition, several other processes that are not obviously dependent on amino acids also stop—for example, the syntheses of rRNA, tRNA, and lipids.

$3'$-Pyrophosphoryl-GDP and -GTP (ppGpp and ppGppp) have been found to accumulate in amino-acid starved strains (Fig. 5.14). The level of these compounds correlates inversely with rRNA and tRNA synthesis—the fastest growing cells have the lowest levels, and the amino-acid starved cells the highest, as if these nucleotide compounds inhibit transcription of the genes for these stable RNAs. It is

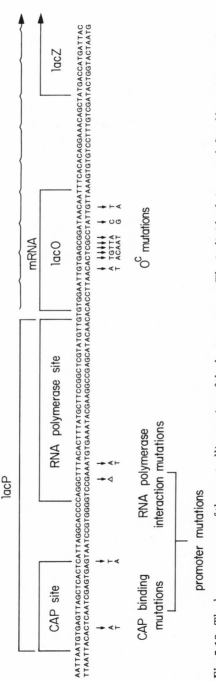

Fig. 5.13 The base sequence of the controlling region of the lactose operon. The individual sites are defined by point mutations (single base-pair changes in the DNA) that affect a given function. Promoter mutations (in either the CAP site or the RNA polymerase site) can increase or decrease the frequency of transcription. Operator mutations (O^c) result in constitutive expression of the operon, because the repressor affinity of the binding site is reduced.

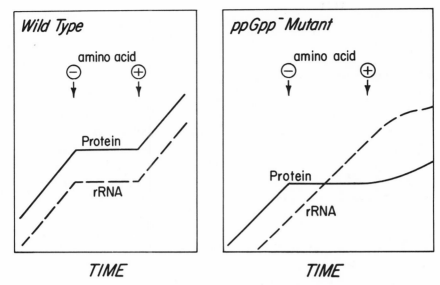

Fig. 5.14 The levels of ppGpp as a general control signal. In wild-type cells amino acid deficiency also brings about the cessation of rRNA synthesis. This linked control is uncoupled in the mutant.

not yet known whether ppGpp interacts with RNA polymerase, another controlling factor, or DNA.

The synthesis of ppGpp requires ribosomes, mRNA, and at least one species of nonacylated tRNA. This seems to be an "idling" reaction occurring on ribosomes when protein synthesis cannot proceed. Mutants that lack the ppGpp-forming enzyme continue to make rRNA during amino acid starvation.

AMMONIA REPRESSION

Another widespread general control mechanism in bacteria concerns metabolic pathways that produce NH_4^+ as a source of nitrogen (for example, deamination of histidine or the fixation of atmospheric nitrogen). Such pathways are not expressed when high levels of NH_4^+ are provided in the medium. The mechanism of control resembles catabolite repression in that the expression of the operons depends on a broad regulatory system as well as on specific induction. Mutations have permitted the detection of two regulatory genes with essential roles in this regulation and three genes whose products have a sharpening effect on the response of the system to changes in the availability of ammonia.

RESPIRATORY REPRESSION

There are other classes of broad control. For example, the levels of a number of cytochromes increase during aerobic growth. By contrast, the levels of certain enzymes in fermentative reactions increase during anaerobic growth. The control of expression of operons involved in respiratory or fermentative processes is still poorly understood. In addition to a general signal reflecting the presence or absence of molecular oxygen, special signals reflecting the levels of certain metabolites seem to be involved. A gene homologous to *cap*, *fnr*, has been identified. The Fnr protein serves as an activator for the expression of genes encoding proteins that function in anaerobic respiration, for example, nitrate reductase, nitrite reductase, and fumarate reductase.

OSMOTIC REGULATION

Increasing the osmotic pressure in the environment can lead to increased synthesis of a K^+-transport system in *E. coli*. The activity of this membrane transport system causes more K^+ to be accumulated in the cytoplasm, thus reducing the osmotic pressure difference across the cell membrane. The nature of the signal is still unknown. (Compare this response to the synthesis of periplasmic oligosaccharides described in Chapter 2.)

Control of Transcription by DNA Rearrangement

An entirely different kind of control involving a semistable mechanism known as **phase variation** can be of importance in determining the pathogenicity of a microorganism. The biological significance of the mechanism can be illustrated by considering the antigenicity of bacterial flagella, a property that allows the host to act defensively by producing specific antibodies to these proteinaceous structures. (This is only one of the components of the host's defense system.) However, certain bacteria, the best characterized being *Salmonella*, have evolved a control mechanism that helps the pathogens to evade the immune response to flagellar proteins. Certain strains can produce two different flagellar antigens, H1 and H2, but only one at a time. The bacteria interconvert at rather high frequency (much higher than can be accounted for by ordinary mutation rates) between the two states.

$$\text{H1 producer} \underset{10^{-4}/\text{generation}}{\overset{10^{-3}/\text{generation}}{\rightleftharpoons}} \text{H2 producer.}$$

Thus, if the host is infected by a population that is predominantly H1 producer, specific anti-H1 immunoglobulins will be produced. However, a significant proportion of the infecting bacterial population can escape this host defense, because those cells only produce H2 antigen. The mechanism for switching back and forth between the H1 and H2 phenotypes involves an inversion of a specific segment of DNA. As illustrated in Figure 5.15, the genes for the H2 antigen and H1 repressor belong to the H2 operon. In phase 1, this operon is not transcribed because of the nonfunctional orientation of its promoter (between sites A and B). Under this condition, operon H1 can be transcribed, resulting in the production of the H1 antigen. When the promoter of the H2 operon is inverted, as shown in phase 2, transcription of this operon occurs. Synthesis of the repressor H1 turns off the transcription of gene H1. Thus only the H2 antigen is synthesized. The inversion of the H2 promoter involves a recombinational event between the two points indicated by the arrows shown in phase 2. This process is catalyzed by an enzyme encoded by the invertible segment itself (not shown).

Exploitation of DNA rearrangement to elude host immune response is not limited to bacteria. Trypanosomes have invented a far more elaborate use of this kind of mechanism to produce a large repertoire of surface antigens, sometimes "outwitting" the antibody system of the host by exhaustion. Genetic rearrangement has also been discovered as a mechanism for activating a pathway for utilizing aromatic β-glucosides by *E. coli*. It has been suggested that this semipermanent control evolved to protect the population from substrates (such as cyanogenic β-glucosides) that would yield toxic products upon hydrolysis. In the presence of such hazardous substrates, strains with the operon rendered inactive by loss of a DNA fragment (an insertion element) will survive. When only utilizable

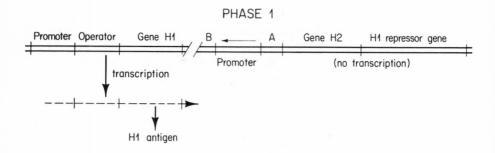

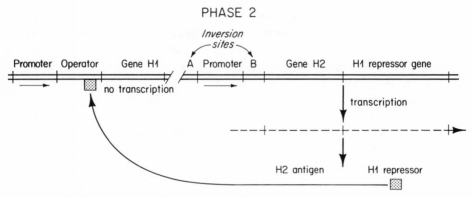

Fig. 5.15 Semipermanent control of gene expression by phase variation (see text).

β-glucosides are present, new progeny in which the operon is active again will emerge.

Regulation at the Level of Translation

GENE SELECTION

As a rule, transcription of an operon is all or none, irrespective of how many structural genes are in the operon, since the process is controlled by the physical state of a single region within which lies the promoter. The same rule applies to the translation of an mRNA encoding a single polypeptide. However, in the case of an mRNA encoding multiple polypeptides, translation in general occurs independently at the beginning of each gene transcript, and the frequency at which translation occurs can vary up to two orders of magnitude. This seems to be largely determined for each gene by a domain of base sequence 5' to the initiation codon, not limited to the ribosome-binding site or the Shine-Dalgarno sequence. Thus the expression of a gene (manifested by the rate of its transcription) is not the only important factor determining the number of its protein products in the cell; the priority of mRNA translation can be of almost equal importance.

REGULATION OF THE FREQUENCY OF TRANSLATIONAL INITIATION

As growth rate increases, so do cell volume and the ratio of ribosomes to genomes. This seems reasonable in view of the obvious connection between the rate of protein synthesis and the concentration of the organelle executing this process. What is more ingenious is that certain ribosomal proteins inhibit the translation of redundant subunits. In *E. coli*, as in many other prokaryotes, the larger 50S subunit has one 23S RNA, one 5S RNA, and 34 proteins (31 different ones, named L1 to L34); the smaller 30S subunit has one 16S RNA and 20 different proteins (named S1 to S20).

Components of the 50S and 30S subunits are assembled independently during the building of ribosomes. Individual proteins of a subunit undergo spontaneous but orderly condensation; some proteins bind directly to the rRNA, while others require the prior binding of certain other proteins. Thus the information for correct assembly of the organelle is self-contained within the three-dimensional structure of the components. Several *in vitro* assembly intermediates were found to resemble those isolated from cells, and the entire process has been observed to go to completion in the test tube. (Mutants defective in subunit assembly accumulate specific incomplete precursors.) Moreover, the proteins that assemble early bind to the 5' end of the rRNA, suggesting that ribosome proteins begin to nucleate around the rRNA still nascent. The basic force involved is no more complex than the correct folding of a protein after the completion of polypeptide synthesis. (The concept of self-assembly will be encountered in a more elaborate form when morphogenesis of bacterial viruses is discussed in Chapter 9.)

The structural genes for the rRNAs are found in a single operon of which there are 7 copies in the *E. coli* chromosome. Transcription of these operons is inhibited either by free ribosomes or by a class of proteins that reflects the concentration of free ribosomes. On the other hand, economical production of a mature ribosome also requires a stoichiometric balance of the numerous components. Thus fine tuning in the synthesis of the various subunit proteins is imperative.

An example of how balanced synthesis of ribosomal proteins is achieved is provided by the discovery of **autogenous inhibition** of its own translation by the L10 protein. This protein impedes not only its own synthesis but also the synthesis of the distally encoded proteins L7 and L12, by acting near the L10 translational start site. It seems that the start sites of the distal genes are occluded by folding of the mRNA and can be "opened" only when a translating ribosome approaches the end of the L10 region (**translational coupling**).

An opposite effect, **translation activation,** is seen in the synthesis of an enzyme that confers drug resistance. When the antibiotic erythromycin combines with the ribosome, mRNA molecules encoding the drug-inactivating enzyme are favored for translation.

Post-Translational Control of Proteins

After the completion of their synthesis, some proteins have to be covalently modified (usually irreversibly) in order to function, others might be covalently modified (usually reversibly) as a way of regulating their function, and still others might be kinetically controlled by effectors (generally small metabolites) that bind reversibly to the protein.

PROTEIN PROCESSING BY COVALENT MODIFICATION

Maturation The making of a final functional protein often is not completed until after the polypeptide is released from the ribosome. In prokaryotes, there are enzymes that trim the formyl or formyl-Met group from the amino terminal end of the polypeptide. Moreover, the side chains of amino acid residues may undergo covalent modification. In *E. coli* the amino-terminal lysine of the S1 ribosomal protein is modified by having its ϵ amino group attached in peptide linkage to an alanine with a methyl group on its α amino group. The functional significance of this modification is unknown. One possibility is that post-translational modification allows bacteria to produce a greater variety of proteins with novel amino acid side chains without expanding their genetic code.

Spatial destination and protein secretion Another class of posttranslational modification determines the spatial status or function of the protein, for instance by covalently joining a protein to another macromolecular structure. Thus the ϵ amino group of lysine in lipoprotein is linked to the terminal carboxyl group of diaminopimelate in murein in the cell wall of gram-negative bacteria.

Post-translational modification plays an even more dynamic role by directing proteins to their proper places, through proteolytic cleavage of precursor polypeptides. Up to 25% of the total protein synthesized by the gram-negative E. coli is not destined for the cytoplasm but is exported to the periplasm or incorporated into the inner or outer membranes. In the case of gram-positive organisms, various enzymes are secreted into the medium. A model has been proposed which offers an explanation for how protein secretion and incorporation into membrane might occur.

According to this model (Fig. 5.16), mRNAs that code for external proteins contain a unique set of codons at the 5' end. Translation of these codons results in a unique amino acid sequence at the amino terminus of the nascent chain. As this "signal sequence" emerges from the ribosomal complex, it triggers attachment of the polysome (a train of ribosomes in the process of translating the mRNA) to the cytoplasmic membrane. This attachment, which may be mediated by specific proteins in the membrane, is probably favored by the hydrophobic nature of the signal sequence. As translation proceeds, the nascent chain is transferred vectorially across the membrane. When translation is finished, the protein is released. The signal sequence usually is not present in the completed protein; proteolytic processing may occur even before synthesis is complete. The net result is that proteins to be exported or secreted are synthesized by cytoplasmic machinery, then translocated out of the cytoplasm. It is not known whether the energy for peptide chain transport across the membrane is provided by the ribosome or by the membrane channel.

This model makes several predictions which have been verified experimentally. For example, proteins known to be secreted or exported are synthesized on membrane-bound ribosomes, and such proteins are made as precursors (peptides with their signal sequences still attached) if the synthesis is carried out in vitro in the absence of

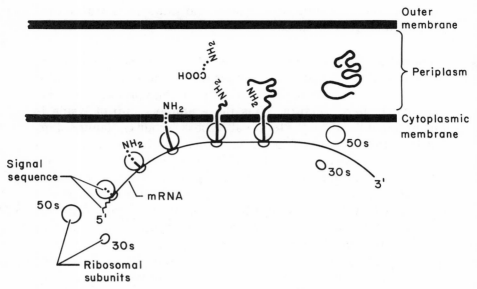

Fig. 5.16 The extrusion of external proteins during their synthesis. The amino terminus of the polypeptide contains a series of amino acids serving as a signal for export by a special set of proteins associated with the inner membrane (not shown). During or after the exit of the polypeptide, the signal sequence is cleaved off. In this scheme the polypeptide is destined for the periplasm. Variations of this mechanism (possibly sharing some processing components in common) are responsible for placing a protein within either the inner or the outer membrane.

processing activity. The model is of general interest because many organisms (even animal cells) may employ a similar mechanism for consigning proteins to various localities. For example, insulin (a secreted protein hormone) is also made in precursor form and may be secreted via an analogous mechanism. An understanding of these processes in bacteria is providing productive models for the elucidation of similar processes in eukaryotic cells and vice versa.

REGULATION OF ENZYME ACTIVITY BY COVALENT MODIFICATION

The scavenging and assimilation of ammonia as a nitrogen source for growth is a reaction of central importance, and is mediated by glutamine synthetase (GS). The reaction is driven by ATP and thus is essentially irreversible:

$$\text{Glutamate} + NH_3 + ATP \xrightarrow{GS} \text{Glutamine} + ADP + P.$$

The complexity of the regulatory mechanism that adjusts the activity of this protein is extraordinary. In particular, covalent modification is brought into play at several levels. Perhaps this elaborate control reflects also the vital importance of keeping in balance the intracellular concentrations of the substrate glutamate and the product glutamine, both of which are necessary for protein synthesis and as participants of other metabolic reactions.

The enzyme GS itself is made up of twelve identical subunits, each of which can be modified by the addition of an AMP to a tyrosine residue. As the adenylylation increases, the catalytic activity diminishes. The culmination of this modification practically results in the abolition of enzyme activity. An adenylyltransferase (ATase) which catalyzes the addition of the AMP group can also remove it. The ratio of these two activities depends on the state of a regulatory protein, P_{II}, to which the enzyme is bound. P_{II} stimulates adenylylation:

$$GS + 12\ ATP \xrightarrow{ATase - P_{II}} GS(AMP)_{12} + 12\ PP.$$

P_{II} itself can be modified by the addition of UMP. When this happens the protein stimulates deadenylylation:

$$GS(AMP)_{12} \xrightarrow{ATase - P_{II}(UMP)_4} GS + 12\ ADP.$$

The interconversion of P_{II} and $P_{II}(UMP)_4$ also appears to be catalyzed by an enzyme, uridylyltransferase (UTase), whose activity is modulated by 2-ketoglutarate and glutamine. UMP addition is activated by 2-ketoglutarate but inhibited by glutamine; conversely UMP removal is activated by glutamine but inhibited by 2-ketoglutarate.

$$P_{II} + 4\ UTP \xrightarrow[\substack{\text{2-Ketoglutarate (activator)}\\ \text{Glutamine (inhibitor)}}]{\text{UTase}} P_{II}(UMP)_4 + 4\ PP$$

$$P_{II}(UMP)_4 \xrightarrow[\substack{\text{Glutamine (activator)}\\ \text{2-Ketoglutarate (inhibitor)}}]{\text{UTase}} P_{II} + 4\ UMP.$$

Consequently, a high glutamine to 2-ketoglutarate ratio, which signals affluence of nitrogenous supply, maintains GS mostly in the inactive form (with AMP) by a cascade of controls. This indirect kinetic control is reinforced by direct feedback inhibition of GS by end-products of pathways to which glutamine contributes nitrogen (see below); these end products include glycine, histidine, tryptophan, glucosamine-6-P, and several nucleotides.

ALLOSTERIC CONTROL OF ENZYME ACTIVITIES

Instantaneous metabolic control is commonly exerted at the level of enzyme activity and prevents excessive chemical flow through specific pathways, while the more gradual controls operating at the levels of transcription and translation prevent wasteful production of proteins. In most cases, kinetic controls do not involve covalent modification. Instead, the activity of the enzyme is influenced by reversible binding of an **effector molecule,** usually a small metabolite. Enzymes that are controlled in this way are typically composed of different polypeptides: catalytic subunits which provide substrate binding sites and regulatory subunits which provide effector binding sites. Since the interaction with the effector at one site can influence the catalytic activity at another site, the term **allosteric** (meaning another space) is used to describe the influence, which can be either positive or negative, depending on the enzyme. In some cases the impact of the effector can be reversed by increasing the concentration of the substrate, in other cases not. Some enzymes have binding sites for more than one kind of effector and are therefore said to be under **multivalent control.**

Feedback inhibition in biosynthesis This kind of kinetic control usually acts on the first enzyme that commits a metabolite to a specific biosynthetic pathway. For example, in the synthesis of histidine, which involves nine enzymes, only the first one (ATP phosphoribosyl transferase) is inhibitable by histidine:

$$\begin{matrix} \text{ATP} \\ + \\ \text{PRPP} \end{matrix} \underset{1}{\to} A \underset{2}{\to} B \underset{3}{\to} C \dots\dots\dots\dots \underset{9}{G \to} \text{histidine.}$$

Another example would be aspartate transcarbamylase, the first enzyme in pyrimidine biosynthesis. Much of what we know about allosteric interactions of proteins is derived from the study of such enzymes. Feedback inhibition is important not only for preventing excessive biosynthesis but in some cases also for allowing the cell to draw preferentially on exogenous end products. The presence of an end product in the environment results in its active transport into the cell, and the increased intracellular concentration of the compound in turn causes repression of the genes specifying the biosynthetic pathway and inhibition of its first enzyme.

Abolition of allosteric control Mutants producing an altered enzyme that is not inhibitable allosterically but still active catalytically may be obtained. These mutants, in contrast to the wild-type strain, may excrete the overproduced end product and are sometimes used for commercial production of important biochemicals (such as essential amino acids and antibiotics).

Feedback inhibition in catabolic, amphibolic, and anaplerotic pathways Feedback inhibition is not only important in preventing redundant biosynthesis but also essential in preventing other kinds of metabolic excesses. An example in controlling a catabolic pathway is the prevention of over consumption of glycerol by feedback inhibition of glycerol kinase, the first enzyme in the pathway. (The rate-controlling step in most other pathways for carbohydrate utilization is apparently at the point of membrane transport.) The key metabolite in glycolysis, fructose-1,6-bisphosphate, is an inhibitor of glycerol kinase (noncompetitive with the substrate glycerol). The importance of this control was revealed by a mutant that produces constitutively a mutationally altered glycerol kinase that is no longer inhibitable by fructose-1,6-bisphosphate; exposure of the cell to glycerol results in overproduction of dihydroxyacetone phosphate, which spontaneously decomposes to the highly chemically reactive methylglyoxal. The result is lethal.

Amphibolic pathways bring about degradation of small molecules as well as provide for the synthesis of essential metabolites. The tricarboxylic acid cycle is an example; it serves both as a way of degrading acetyl CoA for metabolic energy and as a means of generating biosynthetic precursors such as α-ketoglutarate which can be converted to glutamate. The key enzyme mediating the input into the cycle, citrate synthase, is feedback inhibitable by α-ketoglutarate and NADH.

Amphibolic systems like the tricarboxylic acid cycle pose a special problem: depletion of an intermediate, for example α-ketoglutarate, would lead to a shortage of the oxaloacetate needed for reinitiation of a cycle. This problem is averted by the intervention of an **anaplerotic (replenishing) pathway** — in this case the glyoxylate cycle, which produces two molecules of malate from one molecule of isocitrate and one molecule of acetyl CoA. Increasing the concentration of malate leads to the accumulation of oxaloacetate, which in turn can be converted to phosphoenolpyruvate. Phosphoenolpyruvate feedback inhibits isocitric lyase, the enzyme admitting molecules into the glyoxylate cycle.

Questions

5.1. The essentially irreversible biosynthetic pathways for the aspartic acid family of amino acids in *E. coli* is as follows:

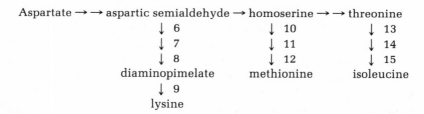

(a) How could one select a mutant requiring both threonine and methionine resulting from a single mutation? Which step(s) would the mutation affect? Would the mutant also have to be fed isoleucine?

(b) Diaminopimelate is found in the mucopeptide of *E. coli*. If lysine-requiring mutants were selected, which step(s) would the mutations be in?

(c) An unusual specific isoleucine-requiring mutant has been isolated; the total pathway sketched above remains normal. What might the mutation affect?

5.2. Four histidine-requiring mutants have been selected. Streaked on minimal medium without histidine, mutant A feeds none of the others, mutant B feeds mutants A, C, and D, and mutants C and D feed only mutant A. (a) How many genes and metabolic steps does the pathway have? (b) How do these data establish the order of the mutational blocks in the biosynthetic pathway? (c) Would the wild-type strain feed any of the mutants?

5.3. Various strains of E. coli show different lag periods when cells growing exponentially on glucose are shifted to lactose as the sole source of carbon and energy. Negligible delay in growth is observed when the reverse shift is carried out. What are the likely reasons?

5.4. After 10^6 cells of a wild-type E. coli strain were plated on agar containing a lactose analog, PG (phenyl-β-D-galactoside), as the sole carbon and energy source, five colonies appeared. When cells of each colony were grown in a liquid medium containing glucose as carbon and energy source and shifted during exponential phase to a fresh medium containing lactose as the carbon and energy source, growth resumed without lag. When 200 wild-type cells were plated on agar containing glycerol and another lactose analog, TONPG (o-nitrophenyl-β-D-thio-galactoside, in which the sugar is linked to the aglycone group by a sulfur instead of an oxygen atom), all of them formed colonies. However, when the PG-utilizing cells were similarly tested, only about 1 in 10^5 cells formed colonies. (a) What might be the nature of the PG-utilizing cells? (b) What might be the nature of their derivatives that can grow on glycerol in the presence of TONPG?

Further Reading

Adhya, S., and S. Garges. 1982. How cyclic AMP and its receptor protein act in Escherichia coli. Cell 29:287–289.

Chock, P. B., S. G. Rhee, and E. R. Stadtman. 1980. Interconvertible enzyme cascades in cellular regulation. Annual Review of Biochemistry 49:813–843.

Crawford, I. P., and G. V. Stauffer. 1980. Regulation of tryptophan biosynthesis. Annual Review of Biochemistry 49:163–195.

Gallant, J. 1979. Stringent control in E. coli. Annual Review of Genetics 13:393–415.

Gausing, K. 1982. Regulation of ribosome synthesis in E. coli. Trends in Biochemical Sciences 7:65–67.

Gottesman, M., A. Oppenheim, and D. Court. 1982. Retroregulation: control of gene expression from sites distal to the gene. Cell 29:727–728.

Gold, L., D. Pribnow, T. Schneder, S. Skinedling, B. S. Singer, and G. Stormo. 1981. Translational initiation in prokaryotes. Annual Review of Microbiology 35:365–403.

Holmes, W. M., T. Platt, and M. Rosenberg. 1983. Termination of transcription in E. coli. Cell 32:1029–1032.

Lee, N. L., W. O. Gielow, and R. G. Wallace. 1981. Mechanism of araC autoregulation and the domains of two overlapping promoters, P_c and

P_{BAD}, in the L-arabinose regulatory region of *Escherichia coli*. *Proceedings of the National Academy of Sciences*, USA 78:752–756.

Lim, L. W., and D. Kennel. 1980. Evidence for random endonucleolytic cleavages between messages in decay of *Escherichia coli* trp mRNA. *Journal of Molecular Biology* 141:227–233.

Lindahl, L., and J. M. Zengel. 1982. Expression of ribosomal genes in bacteria. *Advances in Genetics* 21:53–121.

Losick, R., and M. Chamberlin. 1976. *RNA Polymerase*. New York: Cold Spring Harbor Laboratory.

Losick, R., and J. Pero. 1981. Cascades of sigma factors. *Cell* 25:582–584.

Magasanik, B. 1982. Genetic control of nitrogen assimilation in bacteria. *Annual Review of Genetics* 16:135–168.

Michaelis, S., and J. Beckwith. 1982. Mechanism of incorporation of cell envelope proteins in *Escherichia coli*. *Annual Review of Microbiology* 36:435–465.

Miller, J. H., and W. S. Reznikoff. 1978. *The Operon*. New York: Cold Spring Harbor Laboratory.

Nomura, M., D. Dean, and J. L. Yates. 1982. Feedback regulation of ribosomal protein synthesis in *Escherichia coli*. *Trends in Biochemical Sciences* 7:92–95.

Platt, T. 1981. Termination of transcription and its regulation in the tryptophan operon of *E. coli*. *Cell* 24:10–23.

Pribnow, D. 1975. Nucleotide sequence of an RNA polymerase binding site at an early T7 promoter. *Proceedings of the National Academy of Sciences*, USA 72:784–788.

Reynolds, A. E., J. Felton, and A. Wright. 1981. Insertion of DNA activates the cryptic *bgl* operon in *E. coli* K12. *Nature* 293:625–629.

Silhavy, T. J., S. A. Benson, and S. D. Emr. 1983. Mechanism of protein localization. *Microbiological Reviews* 47:285–312.

Smith, G. R. 1981. DNA supercoiling: another level for regulating gene expression. *Cell* 24:599–600.

Takeda, Y., D. H. Ohlendorf, W. F. Anderson, and B. W. Mathews. 1983. DNA-binding proteins. *Science* 221:1020–1025.

Wittmann, H. G. 1982. Components of bacterial ribosomes. *Annual Review of Biochemistry* 51:155–183.

———— 1983. Architecture of prokaryotic ribosomes. *Annual Review of Biochemistry* 52:35–65.

Yanofsky, C. 1981. Attenuation in the control of expression of bacterial operons. *Nature* 289:751–758.

Yura, T., and A. Ishihama. 1979. Genetics of bacterial RNA polymerases. *Annual Review of Genetics* 13:59–97.

6

DNA Replication and Mutation in Bacteria

ALL THE basic information of the bacterial genome is contained in a single chromosome which is a circular DNA molecule. Higher organisms, which usually inherit a pair of homologous chromosomes (one from each parent, so that the two are not identical), are said to be **diploid;** bacteria, because they have a single chromosome, are said to be **haploid.** In addition to genes, the bacterial chromosome, like chromosomes of all other organisms, contains sites that regulate gene expression, such as promoters, operators, CAP-binding sites, and attenuators (see Chapter 5).

General Features of Bacterial Replication

The circular *E. coli* chromosome is conventionally partitioned into 100 equal sections numbered in clockwise order as "minutes" (because it takes about one minute to transfer a section of the chromosome into the recipient cell during conjugation; see Chapter 7). A round of chromosome replication begins with the **initiation** of DNA synthesis at a specific site called *ori*, for origin of replication. This site is located at minute 80. Initiation of a new round of replication is a complex process that is only vaguely understood. It is known that the *ori* site is acted upon by specific proteins, some of which function only during initiation. Near *ori* are structural genes whose products aid the initiation process. As shown in Figure 6.1, initiation involves opening the double-stranded DNA molecule at *ori* and laying down a

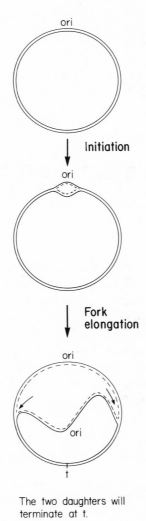

ori

Initiation

ori

Fork
elongation

ori

ori

t

The two daughters will
terminate at t.

Fig. 6.1 Bidirectional replication of the *E. coli* chromosome initiated at the *ori* site.

primer on each side, which is a polynucleotide with base sequence complementary to the template in the *ori* region (adenine paired with thymine, or uracil, as the case may be; and guanine paired with cytosine). The primer is composed of RNA and its synthesis is directed by RNA polymerase.

Chromosome replication is a regulated process that is synchronized closely with cell division. The regulatory factors involved act at the initiation step. For example, cells growing in rich broth at 37°C divide once every half hour; correspondingly, initiation of DNA replication occurs every half hour. By contrast, in acetate mineral medium both events occur once in 5 hours.

After the *ori* region is primed, DNA polymerases and associated proteins take over and synthesis proceeds with both strands serving as template. Because the site of new DNA synthesis appears like a "Y" made up of double-stranded DNA, the structure is called a fork and the process of polymerization is called **fork elongation.** The rate of fork movement is constant, independent of cell division time. Once the forks begin traversing from *ori*, about an hour is needed to replicate the entire genome. The bacterial chromosome is said to have **bidirectional** replication because two forks are initiated at *ori*, and the replicating complexes travel in opposite directions. The intermediate of DNA replication is called a **theta structure** because of its characteristic profile (see the bottom drawing in Fig. 6.1). There appears to be a specific region, 180° from *ori*, where the two forks will converge to complete replication. The two complete daughter chromosomes will then be separated and partitioned into their respective cells.

Errors in Replication

The process of replication is remarkably accurate, considering the number of nucleotides that must be properly matched during the elongation of a daughter strand of DNA. But because mistakes do occur, organisms have evolved mechanisms for proofreading and making corrections. Nonetheless, not all errors are corrected and the mistakes are perpetuated in the progeny DNA as mutations. A **mutation** may be defined as a stable and heritable change in the sequence or number of nucleotides in a nucleic acid molecule. A bacterial culture will always contain a low proportion of spontaneous mutants of various kinds. The probability of a base-pair change introduced during replication at a particular position of the chromosome may vary roughly from 10^{-7} to 10^{-11}, because mutability is dependent upon the sequence surrounding the base pair (that is, it is context-dependent) and on whether a base is methylated or not (see "DNA Repair," below). In addition to intrinsic errors in DNA replication, alterations may also be introduced by the action of mutagens in the organism's environment. It should be borne in mind that although some mutations might be disastrous for the progeny of a cell, a certain frequency of random mutations is necessary if the species is to have enough genetic variety to respond to selective pressures in the environment.

The "normal" standard against which a mutant is compared is called the **wild-type** strain. The term originally referred to the organism as found in nature (in the wild) but is now loosely applied to the strain used in the laboratory from which mutant stocks are derived. Mutants can be differentiated from the wild-type strain by chemical sequencing of DNA or by other direct physical methods, but in practice mutations are usually recognized by their effects on the physiology or metabolism of the cell, that is, their **phenotype.**

Mutations that block degradative or biosynthetic pathways are not necessarily lethal to the cell. For example, a mutation in the lactose system that abolishes the organism's ability to degrade lactose would not prevent growth unless lactose were the only carbon or energy source available. Similarly, a mutation in the histidine system which abolishes the biosynthesis of histidine would not prevent growth as long as histidine was available in the medium. These mutations are sometimes referred to as "nonlethal." On the other hand, mutations that interfere with essential functions—such as the ability to make RNA polymerase—cannot be overcome by nutritional manipulations. These mutations are generally referred to as "lethal."

Some mutations in essential functions merely restrict the conditions under which an organism can grow. For example, mutations can occur in a gene which cause the synthesis of an altered protein that is stable only at low temperatures. Heat lability is generally caused by amino acid substitutions in the protein. Whereas wild-type *E. coli* grows in the range of 20–45°C, a mutant in RNA polymerase might be found among cells that can grow at 30°C but not at 40°C. In this case the lower temperature range would be the **permissive** condition and such a mutant would be called **conditional.** Cold-sensitive mutants, mutants that are able to grow only in media of high osmotic strength, and many other conditional mutants are known.

There is also a class of mutants in essential functions that does not need special conditions for growth. The mutant enzyme retains its basic function but is altered in an easily recognizable way, such as in its sensitivity to an inhibitor. For example, rifamicin-resistant mutants contain an altered polypeptide subunit in RNA polymerase.

SELECTING MUTANTS IN THE LABORATORY

As we saw in Chapter 5, wild-type *E. coli* is able to use lactose as a carbon source because it possesses *lacY* and *lacZ* genes. Mutation in either gene may result in a strain unable to grow on lactose (that is, one that has a Lac⁻ phenotype). The spontaneous frequency of such mutants in a culture of the wild-type strain is roughly 10^{-5}. To find one mutant cell amid 10^5 wild-type cells, one could directly screen colonies by using a lactose indicator plate; that is, one could look at 1,000 plates with 1,000 colonies on each and expect to find about 10 *lac⁻* mutants. To avoid such a laborious process, one often uses some kind of selection that favors growth of the rare mutants. A common technique is to treat the culture with an agent that kills the unwanted *lac⁺* cells but not the *lac⁻* cells.

To take an example, a large population containing one *lac⁻* mutant in 10^5 cells can be placed in a minimal medium with lactose as sole carbon source and treated with penicillin, known to kill only growing cells. Since the wild-type bacteria (*lac⁺*) are the only ones able to grow on lactose, they would be killed by the penicillin, thus increasing the frequency of *lac⁻* mutants in the surviving population. Even easier methods are often available in bacterial genetics. For selecting *lac⁻* mutants, a toxic lactose analog (o-nitrophenyl-β-D-thiogalactoside) can be included in the medium with an inducer and an appropriate carbon and energy source; only *lac⁻* (permease-negative) mutants will grow. The growth of the wild-type cells will be inhibited. One could put 10^8 cells on a single plate and directly obtain the 10^3 or so mutant clones.

Classes of Mutations

Mutations range from the simplest change in one base pair to additions, deletions, or rearrangements in the sequence of thousands of base pairs. These changes can result from errors of DNA replication, damages caused by chemical or physical agents, or special events promoted by genetic elements. The following discussions of mutagenesis, reversion, and suppression apply to both cellular and viral genomes.

BASE SUBSTITUTION

During DNA replication, a wrong base might be incorporated into a daughter strand in violation of the normal rules of base pairing. For example, an A, G, or C might be mistakenly incorporated into a nascent strand at a position calling for a T (because the template contains an A). If a C is incorporated into the newly synthesized strand opposite the A, in the next round of replication the C will direct the incorporation of a G. Thus, an original wild-type AT pair will be converted to a mutant GC pair in one of the daughter chromosomes. On the other hand, if an A or G is incorporated, the result would be a change from an AT to a TA or CG pair, respectively. A change involving the substitution of one purine for another purine (changing an A to G, or vice versa) or one pyrimidine for another pyrimidine (T to C, or vice versa) is called a **transition mutation.** A change involving the substitution of a purine for a pyrimidine (A to T or C; G to T or C), or vice versa, is called a **transversion mutation.** Since there are just four different pairs (AT, TA, GC, CG), only three base-pair substitutions are possible at a given position.

When an error is introduced into a newly synthesized strand, the new DNA duplex becomes locally **heterozygous.** The parental strand will give rise to a wild-type DNA duplex, while the daughter strand will give rise to a mutant DNA duplex. The wild-type and mutant duplexes will be inherited by different progeny of the cell. One or more cell divisions may be required for the phenotype of the mutation to be manifested. This in turn may depend on whether or not the error was originally introduced into the sense strand (in the case of a coding region), on the number of copies of the chromosome present in the cell, and on whether the mutation is **dominant** (whose phenotype can be expressed even in the presence of a normal gene) or **recessive** (whose phenotype is not expressed in the presence of a normal gene). The delay in the manifestation of the effect of a mutation is called **phenotypic lag.**

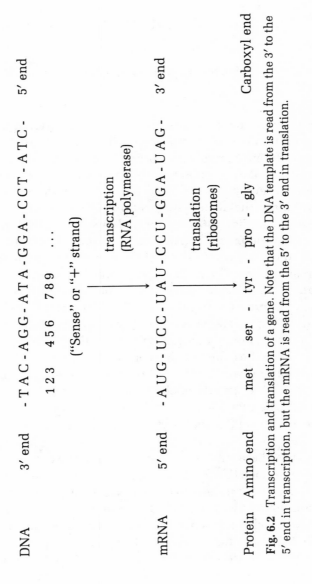

DNA 3′ end - T A C - A G G - A T A - G G A - C C T - A T C - 5′ end

123 456 789 ...

("Sense" or "+" strand)

transcription
(RNA polymerase)

mRNA 5′ end - A U G - U C C - U A U - C C U - G G A - U A G - 3′ end

translation
(ribosomes)

Protein Amino end met - ser - tyr - pro - gly Carboxyl end

Fig. 6.2 Transcription and translation of a gene. Note that the DNA template is read from the 3′ to the 5′ end in transcription, but the mRNA is read from the 5′ to the 3′ end in translation.

DNA (wild type)　- T A C - A G G - A T A - G G A - C C T - A T C -

DNA (mutant)　- T A C - A G G - A $\overset{*}{C}$ A - G G A - C C T - A T C -

mRNA　- A U G - U C C - U $\overset{*}{G}$ U - C C U - G G A - U A G -

Protein　- met - ser - cys - pro - gly

Fig. 6.3　A transition mutation changing tyrosine to cysteine.

Figures 6.2 and 6.3 together illustrate the process whereby a single base substitution can lead to the synthesis of an altered protein containing a different amino acid. In Figure 6.2 and similar subsequent figures the DNA template strand (the sense or + strand) is shown with its 3' end on the left, and the mRNA with its 3' end on the right. Mutations that lead to single amino acid substitutions such as the one shown in Figure 6.3 are termed **missense mutations.**

Within a given **codon** (a group of three nucleotides that specifies a particular amino acid; see Table 6.1) there are nine possible base substitution mutations. The mutations possible in the third codon in Figure 6.2 are listed in Table 6.2. Note that not all of them necessarily affect the protein product. In this case, one of the base substitution mutations (ATG) still retains tyrosine in the polypeptide (because there is more than one codon for tyrosine and most other amino acids). These would be called phenotypically "silent" mutations. Six of the nine mutations would replace tyrosine by another amino acid (asparagine, histidine, aspartate, phenylalanine, serine, or cysteine). These might also be silent — it would depend on how critical the side chain of the amino acid at that particular position is for the proper function of the protein.

Two of the nine base substitution mutations in the example result in **termination codons** (UAA and UAG; the other termination codon is UGA); polypeptide synthesis in such a case is shown in Figure 6.4. Any mutation that converts a codon for an amino acid to a termination codon is called a **nonsense mutation.** This kind of mutation generally abolishes the function of the protein.

When base substitution mutations occur in regions of DNA not specifying proteins (promoter, for example), it is difficult to guess their effect. Depending on the function of that DNA, certain base substitution mutations may have profound effects.

Table 6.1.   The genetic code, whereby a sequence of three bases in mRNA (a codon), read from the 5′ to the 3′ end, specifies an amino acid.

First position (5′ end)	Second position				Third position (3′ end)
	U	C	A	G	
U	Phe	Ser	Tyr	Cys	U
	Phe	Ser	Tyr	Cys	C
	Leu	Ser	Term	Term	A
	Leu	Ser	Term	Trp	G
C	Leu	Pro	His	Arg	U
	Leu	Pro	His	Arg	C
	Leu	Pro	GluN	Arg	A
	Leu	Pro	GluN	Arg	G
A	Ileu	Thr	AspN	Ser	U
	Ileu	Thr	AspN	Ser	C
	Ileu	Thr	Lys	Arg	A
	Met	Thr	Lys	Arg	G
G	Val	Ala	Asp	Gly	U
	Val	Ala	Asp	Gly	C
	Val	Ala	Glu	Gly	A
	Val	Ala	Glu	Gly	G

Ala, alanine	Gly, glycine	Pro, proline
Arg, arginine	His, histidine	Ser, serine
Asp, aspartate	Ileu, isoleucine	Term, termination codon
AspN, asparagine	Leu, leucine	Thr, threonine
Cys, cysteine	Lys, lysine	Trp, tryptophan
Glu, glutamate	Met, methionine	Tyr, tyrosine
GluN, glutamine	Phe, phenylalanine	Val, valine

Table 6.2.   Possible consequences of single base changes of the DNA triplet ATA.

Wild-type triplet	Mutant triplet	mRNA codon	Amino acid
ATA (tyr)	TTA	AAU	asn
	GTA	CAU	his
	CTA	GAU	asp
	AAA	UUU	phe
	AGA	UCU	ser
	ACA	UGU	cys
	ATT	UAA	term
	ATG	UAC	tyr
	ATC	UAG	term

DNA (wild-type)   - T A C - A G G - A T A - G G A - C C T - A T C -

DNA (mutant)   - T A C - A G G - A T $\overset{*}{C}$ - G G A - C C T - A T C -

mRNA   - A U G - U C C - U A $\overset{*}{G}$ - C C U - G G A - U A G -

Protein   met - ser - STOP

Fig. 6.4 A transversion mutation giving rise to a termination codon.

DELETIONS, FRAME SHIFTS, INSERTIONS, AND INVERSIONS

A **deletion** removes one or more base pairs, and extensive ones can result in the disappearance of entire genes. Deletions generally eliminate gene function. On the other hand, the **addition** of one or more base pairs can also happen. The bases added may be a reiteration of a neighboring sequence, sometimes sufficiently long to include entire genes (gene amplification or duplication). Additions may also be the result of insertion of a sequence with special genetic activities (see below).

A **frame-shift mutation** involves deletion or addition of a small number of base pairs (but not a multiple of 3) which shifts the reading frame in translation. Such mutations dramatically alter the protein product (Fig. 6.5) by creating a new amino acid sequence from that point on. Usually, the translation is terminated prematurely, since within the region of the new reading frame a nonsense codon is likely to be encountered.

Spontaneous gene-inactivating mutations are frequently caused by **insertion sequences** (IS) that range from 800 to 1,500 base pairs. Insertion sequences are discrete DNA sequences normally occurring in the bacterial chromosome that have the ability to transpose, that is, to cause the appearance of the same sequence at new genetic sites. Insertion sequences have no intrinsic phenotype, but can inactivate the genes into which they are inserted. Mutations induced by these sequences are usually highly polar (see below). Transposable elements are known to exist also in the genomes of higher organisms.

Because of the proclivity of transposable elements to multiply in copy number in a genome without conferring obvious advantage to the host, they are considered by some students of evolution as **selfish genes.** The transposable elements are capable of inducing deletions and **inversions** (the flipping of DNA sequences) in addition to insertion mutations.

DNA (wild-type) - T A C - A G G - $\overset{*}{A}$ T A - G G A - C C T - A T C -

delete A

DNA (mutant) - T A C - A G G - TA - G GA - C CT - A TC - X -

mRNA - A U G - U C C - AU - C CU - G GA - U AG - Y -

new reading frame

Protein   met - ser - $\overset{*}{\text{ile}}$ - $\overset{*}{\text{leu}}$ - $\overset{*}{\text{asp}}$ -

Fig. 6.5 A "minus one" frame-shift mutation.

POLAR MUTATIONS

A single mutation in a multigenic operon may cause the loss of more than one protein. For example, in Figure 6.6 a nonsense or frameshift mutation in the promoter-proximal gene A will result in a lower level of expression of the distal genes B and C. The nearer the mutation in gene A is to the promoter, the stronger its effects. These mutations are known as **polar mutations,** since they only affect the expression of other genes in a direction away from the promoter. Insertion of a transposable element usually has a polar effect.

Since each of the gene transcripts in the mRNA of an operon is translated by ribosomes which attach and initiate independently at individual ribosome-binding sites, it is not the mere stopping of ribosomal progress in the mutated proximal gene that inhibits distal gene expression. Rather, it appears that within operons there are ρ-dependent transcription termination sites that are effective to varying degrees when translation within one gene stops. The frequency of the ρ-dependent transcription termination depends on the distance of the mutation site from the termination site. Part of the evidence for this model comes from the finding that certain mutations in the ρ gene suppress polarity.

The Action of Mutagens

Although rare spontaneous mutants can be recovered with appropriate selective procedures, mutagens are often used so that the propor-

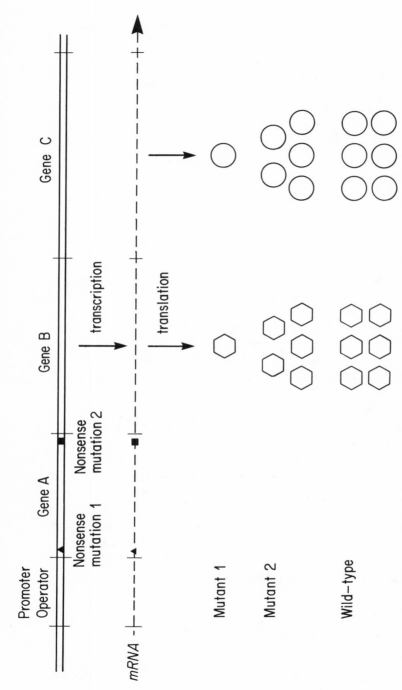

Fig. 6.6 Polarity effects of nonsense mutations in a multigenic operon. Nonsense mutant 1 produces greatly diminished amounts of proteins B and C because the relatively long stretch between the mutation and gene B increases the probability of transcription termination of the mRNA before gene B. In contrast, mutant 2 produces almost normal amounts of proteins B and C because there is little separation between the mutation and the next gene (premature transcripts not shown). By the same principle nonsense mutations in gene B will affect the level of protein C but not protein A.

tion of mutants in a population will become high enough for direct screening. The mechanisms of mutagenesis are often complicated, even though the end result may be as simple as a base-pair substitution. Mutagens do not act by directly converting one base pair into another; intermediates of altered DNA are generally involved in the process of mutagenesis. Sometimes normal replication of an abnormal template fixes the mutant DNA sequence, and sometimes an error in DNA repair is responsible for the mutated sequence. Mutagenesis is important, not only because of its central role in the study of genetics, but also because of the implication of environmental mutagens in the development of certain kinds of cancer.

Direct change of one base into another   Hydroxylamine can specifically react with a cytosine residue in one strand, converting the C to a derivative with base-pairing properties like thymine. The eventual result will be the conversion of a CG pair to a TA pair, thus producing a transition mutation.

Base analogs   Base analogs resemble normal bases, and are utilized by cells in the same way exogenous bases are used: by uptake and conversion into the nucleotide. Bromouracil (Bu), for example, will normally act like thymine in base pairing, and is thus incorporated opposite A during DNA replication. However, when Bu is in its rare tautomeric form, it acts like cytosine in base pairing. Thus, DNA *already containing Bu* may occasionally direct the laying down of a G during replication, forming a BuG pair. In the next replication, this G will pair with C, thus converting the original TA (via BuA) to a CG. On the other hand, if Bu were *incorporated* in its rare form (Fig. 6.7), it would pair with G instead of A. In that case, the net process would convert a CG pair (via BuG) into a TA pair. In both cases, a transition mutation is induced.

Ultraviolet light and X ray   Among other effects, ultraviolet light causes the formation of thymine dimers between adjacent Ts on a DNA strand. This prevents normal DNA replication and would be lethal, were there no repair systems, of which E. coli has at least four (see "DNA Repair," below). However, the repair processes are not foolproof, and the mistakes remaining or made during the processes are manifested as mutations which are mostly base substitutions

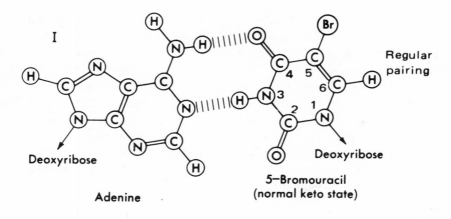

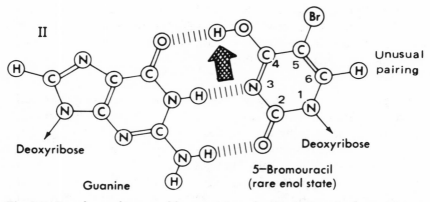

Fig. 6.7 Regular and unusual base pairing of 5-Bu. (I) Regular base pairing (in the common keto form) with adenine. (II) Base pairing (in rare enol form) with guanine. The heavy arrow in II indicates the displacement of the proton in the tautomerization of 5-Bu. (From B. D. Davis, R. Dulbecco, H. N. Eisen, and H. S. Ginsberg. 1980. *Microbiology*. 3rd ed. New York: Harper and Row, p. 202.)

(both transitions and transversions), sometimes affecting two consecutive bases. Frame shifts and deletions also result. X rays and ionizing radiation cause strands to break and purine and pyrimidine rings to open.

Frame-shift mutagenesis Planar aromatic compounds like acridines are thought to stack or intercalate between bases, distorting the DNA helix and thus permitting illegitimate pairing. As a result of this distortion, replication yields two strands of unequal length: the shorter strand may have sustained a deletion or the longer strand may have acquired additional bases.

Alkylating and cross-linking agents Ethylmethane sulfonate reacts mainly with guanine and cytosine, changing their pairing properties in further replication. Nitrosoguanidine gives rise in the cell to an alkylating agent. This very potent mutagen tends to attack at the replicating fork of DNA, which accounts for occasional clustered multiple mutations. Mustard gas causes chemical cross-links between adjacent purine bases.

Mutator genes *E. coli* has a number of genes, the products of which function to repair aberrant DNA. Mutational alteration of any of these "mutator" genes yields strains that have higher spontaneous mutation frequencies. In addition, mutant DNA polymerases that increase or decrease mutation frequency have been found.

Forward and Back Mutation (Reversion)

Forward mutation is the conversion of a gene from a generally functional wild-type to a different and often nonfunctional state. A **back mutation** or reversion is just the opposite: it converts the altered gene to its original state. The frequency of back mutation is much lower than that of forward mutation, because there are many more ways to change a gene than ways to restore it. Figure 6.2 shows that the T in the third triplet (ATA) is only one of hundreds of bases in the gene within which a mutation could occur; in contrast, once that particular mutation has occurred, a reversion to the wild-type codon would be a much rarer event.

The frequency of back mutation, just like that of forward mutation, can be increased by mutagenesis. However, a mutagen that acts in a very specific way may not be able to induce a strict reversion of a mutation caused by that same agent. For instance, hydroxylamine can change a CG base pair to a TA base pair by attacking the cytosine. But the TA cannot be changed back to CG by hydroxylamine, because the mutagen reacts *only* with cytosine. For a similar reason, mutagens causing transitions do not reverse frame-shift mutations, but mutagens that cause frame shifts generally do. Deletions involving more than one base pair generally do not revert at all. Mutations caused by an insertion element in general revert only by exact excision at very low frequency; excision is not effected by any known mutagen.

SUPPRESSION

In the section above, only strict back mutation to wild-type sequence was considered. However, other alterations of base pairs away from the mutational site might also restore function without producing the original base-pair sequence. Consider, for example, the case of a mutation in the structural gene of tryptophan synthetase affecting the protein at position A23 (Table 6.3). The mRNA of the wild-type sequence is GGA (codon for glycine); the corresponding sequence in the mutant is AGA (arginine). A molecule of tryptophan synthetase with an arginine in that position is not enzymatically active, and consequently the cell will require tryptophan for growth. Among the

Table 6.3. Revertants of tryptophan synthetase mutant A23. A certain glycine codon (GGA) in the wild-type gene was mutated to an arginine codon (AGA) resulting in a tryptophan requirement for growth. Phenotypic revertants were then examined to see which of the nine possible changes could restore an enzymatically active protein.

Codons derivable from AGA by a single base change	Triplet code for	Type of change	Found among revertants
GGA	Gly	Transition	Yes[a]
UGA	Term	Transversion	No
CGA	Arg	Transversion	No
AAA	Lys	Transition	No
AUA	Ileu	Transversion	Yes[b]
ACA	Thr	Transversion	Yes[c]
AGG	Arg	Transition	No
AGU	Ser	Transversion	Yes[b]
AGC	Ser	Transversion	Yes[b]

a. Spontaneous and 2-aminopurine induced.
b. Spontaneous.
c. Spontaneous and ethylmethane-sulfonate induced.

nine possible single base changes of the AGA triplet, one is the original GGA (glycine). However, this was not the only back mutant obtained. Phenotypic revertants (pseudo-revertants) were also found with isoleucine (AUA), threonine (ACA), and serine (AGU and AGC) at position A23 in the enzyme. Thus, four of the other base substitutions of the triplet gave acceptable amino acids. Of the other possible substitutions, two, AGG and CGA, also code for arginine (like the mutant itself); another, AAA, gives lysine, which is unacceptable; and the remaining UGA is a nonsense triplet.

Reverse mutations giving rise to triplets that differ from the original wild-type triplet but allow the protein to be functional are called **suppressor mutations** — secondary mutations that correct the effect of the primary mutation. These particular suppressor mutations are **intragenic** and **intracodonic** (in the same gene and codon as the original mutation). One could also find intragenic suppressor mutations that are not in the same codon. For example, an amino acid substitution at a nearby position in the three-dimensional structure of the protein — perhaps an arginine to glycine switch restoring the original net charge — might again give a functional enzyme.

Suppressor mutations lying outside the original gene — **extragenic suppressors** — are also possible. An example would be when a gene coding for another protein could mutate to give a new protein that assumed the activity of the lost one. However, the best studied class is the extragenic suppressors of nonsense mutations, where the suppressor mutation causes an alteration in a tRNA so that it will read a nonsense codon and insert a functionally acceptable amino acid at that position. In the example given in Figure 6.8, the suppressor mutation is in the triplet that specifies the anticodon itself, but there are other possibilities. Although the kinds of nonsense suppressor mutations are limited, the action of these suppressors is not limited to nonsense mutations in any particular gene.

Altered tRNAs are not the only kind of extragenic nonsense suppressor. Another class involves particular alterations in ribosome structural proteins, causing distortion and occasional mistranslation of a nonsense codon as a "sense" codon. Certain ribosome mutations increase the frequency of mistakes in translation over a broad range of codons and not just for nonsense codons. There also exist mutations of opposite effect.

Suppressor mutations also occur for frame-shift mutations. Here, the most common case is intragenic suppression of the frame shift by

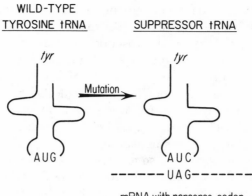

WILD-TYPE
TYROSINE tRNA

SUPPRESSOR tRNA

mRNA with nonsense codon

Fig. 6.8 Nonsense suppression of UAG by a mutation in the third position of the anticodon of a tRNA for tyrosine. There are a major and a minor tyrosine tRNA (more and less abundant in the cell respectively), both bearing the 3'-AUG-5' anticodon. It is the minor tRNA that is usually involved in suppression, since dispensing with it would have no apparent effect on the rate of translation of the UAC codons in mRNAs.

a second mutation in the same gene restoring the reading frame (for example, a -1 suppressed by a $+1$ nearby; see Fig. 6.9). In this example the protein from the suppressed strain has an isoleucine in place of a tyrosine (compare with Fig. 6.2):

DNA (wild-type) -TAC-AGG-$\overset{*}{A}$TA-GGA-CCT-ATC-

↓ delete A

DNA (frame shift mutant) -TAC-AGG- $\underline{TA}$-$\underline{GGA}$-$\underline{C}$ $\underline{CT}$-$\underline{A}$ $\underline{TC}$-

new reading frame

↓ insert G

DNA ("revertant") -TAC-AGG-$\underline{TA}\overset{*}{\underline{G}}$-$\underline{GGA}$-$\underline{CCT}$-$\underline{ATC}$-

reading frame restored

mRNA -AUG-UCC-AUC-CCU-GGA-UAG-

Protein met - ser - $\overset{*}{i}$le - pro - gly

Fig. 6.9 Intragenic suppression of a frame-shift mutant.

THE AMES TEST

When the nature of the original mutation is known, selection of back mutants is an excellent way to test for the presence and nature of mutagens in samples. In the Ames test, a set of several characterized his^- mutants (including base-pair substitutions and frame shifts) is used. The sample to be assayed for mutagenic activity is simply mixed with the mutants and spread on a minimal plate without histidine. Samples containing mutagens induce reversions that permit the cells to grow into colonies. The plate test may also be modified by including some liver extract and an NADPH-generating system; this allows hydroxylation of certain compounds, which then become mutagenic. A number of compounds are known to become carcinogenic only after they are similarly hydroxylated *in vivo*. The majority of proven carcinogens act as mutagens in this test.

DNA Repair

Of such paramount importance are the protection and the faithful reproduction of the genome that studies on *E. coli* have already identified some 30 genes whose products are involved in different processes that repair damaged DNA. Alterations in any one of these genes can alter the mutation frequency.

The information encoded in DNA can be deranged by a variety of causes of both intrinsic and extrinsic origin. Intrinsic errors can arise from:

(1) mismatched base pairs that arise during DNA replication;
(2) spontaneous deamination of cytosine (producing uracil, which will change a CG pair to a TA pair in a subsequent round of replication), adenine (producing hypoxanthine, which may pair with cytosine and thus convert an AT to a GC pair), and guanine (producing xanthine, a noncoding base);
(3) spontaneous methylation of bases by S-adenosylmethionine at unusual positions, which can interfere with proper base pairing;
(4) spontaneous depurination caused by heat and catalyzed by acid (at 37°C an *E. coli* chromosome loses about 10 purines per day and 0.1 cytosines per day by spontaneous deamination); and

(5) gross genetic rearrangements that involve breaking and rejoining phosphodiester bonds.

Extrinsic errors are caused by physical and chemical mutagenic agents. With very few exceptions (such as double-strand breaks separated by only a few bases), all of the known damages to DNA are correctable by special enzymes, and some of the mechanisms have been identified. Repair mechanisms almost always take advantage of the fact that DNA is double stranded; thus the information on the undamaged strand is used as a blueprint to guide restorative synthesis after the damaged parts are removed.

ERROR REMOVAL BY PROOFREADING DURING SYNTHESIS OR BY EXCISION AND RESYNTHESIS

There are at least two known ways by which incorrect bases are detected and removed from DNA. The first involves the DNA polymerases. During replication there is a pause after the addition of each base; if the base is incorrectly paired to the template strand (about once in every 1000 bases), then a special $3'$-$5'$ exonuclease activity will excise the base. If the proofreading system fails, a mismatched base pair can then be detected by a complex enzyme system (known as the mismatch-repair system) that specifically recognizes the mismatched base on the newly synthesized strand. The repair entails incising the strand close to the $5'$ side of the lesion, cleaving a small number of nucleotides, and thus removing the offending unit. The gap is filled by additional replication and ligation.

Illegitimate bases or remnants of bases like uracil, hypoxanthine, 3-methyladenine, and purine and pyrimidine rings opened or fragmented by ionizing radiation or by high pH are recognized and removed by N-glycosylases without hydrolyzing phosphodiester bonds. At least seven different enzymes of this kind with various specificities are present in E. coli, and some of these are inducible by the damaged DNA. Single base gaps are also generated by spontaneous depurination. Irrespective of their origin, gaps can be recognized by one or more endonucleases and the missing base replaced by the cut-and-patch process. DNA damaged by cross-linking of bases on opposite strands can also be repaired by a similar process, one strand at a time, provided the two bases are not paired partners but occupy staggered positions.

ERROR-FREE REPAIR BY TRANSMETHYLASE

Methylation or ethylation of guanine to give O^6-alkylguanine can be repaired by a special enzyme that catalyzes the transfer of the alkyl group onto itself. The enzyme commits "altruistic suicide" by this action because the alkyl group transferred to the cysteine residue of the protein inactivates the enzyme irreversibly. An *E. coli* cell normally contains about 20 molecules of this enzyme, but after exposure to an alkylating agent the number can rise to 3000 to 10,000. Thus cells can acquire short term "immunity" after an exposure to the mutagen.

Ironically, by self-methylating certain cytosines in the DNA for protection against specific endonucleases, the cell also creates a highly mutable base. This is because spontaneous deamination of the 5-methylcytosine gives thymine, a natural base. The event escapes censorship, since there is no way to determine whether the T or the G at the mismatched locus is in error. If the T is not excised, the consequence will be the conversion of a CG to a TA pair.

PATHWAYS FOR REPAIR OF UV DAMAGE

Sunlight is so ubiquitous and destructive of DNA that no known bacteria inhabiting the surface of the earth are without special mechanisms to repair radiation damage. Likely exceptions are the thermophilic bacteria that live close to the volcanic vents at the bottom of oceans, but those cells might have more efficient ways of coping with the rapid rate of depurination caused by the high ambient temperature (well over 300°C). Four different systems in *E. coli* are known to participate in repairing the DNA damaged by ultraviolet irradiation, of which the formation of thymine dimer is the principal lesion. Synthesis of DNA will stall in front of the dimer until the damage is repaired, thus avoiding erroneous copying.

Photoreactivation This is one repair mechanism that restores the original bases without relying on the complementary DNA strand. The energy of visible light (340 to 400 nm) can be utilized by an enzyme to dissociate a covalently linked pyrimidine dimer (cyclobutylpyrimidine) and restore the original state of the thymines. Photoreactivating enzymes are found in many kinds of cells, ranging from mycoplasmas (whose genomes are about 20% as large as that of *E. coli*) to simple eukaryotes.

In addition to this **light repair** mechanism, there are three other processes that fall under the category of **dark repair.**

Nucleotide excision repair In *E. coli* there are at least five different incision endonucleases recognizing damaged strands of DNA. A complex nuclease encoded by the *uvrA,B,C* genes can detect DNA abnormalities, such as the presence of pyrimidine dimers and polycyclic hydrocarbon adducts, and can incise the phosphate backbone near the lesion. The enzyme is especially active at sites adjacent to thymine dimers and is induced following damage of DNA by ultraviolet light. The exposed 5′ phosphate end of the single strand bearing the lesion becomes a signal for exonuclease to remove the lesion. The gap is then filled by DNA polymerase and the ends joined by DNA ligase.

Sister-strand exchange by post-replicative recombination A way to bypass the blockade imposed on the DNA polymerase by the lesion in the damaged template strand is to make a new start beyond the lesion. After the complementary region of the normal sister strand is replicated, an exact portion of that template is resected and grafted to fill the gap created by the bypass in the strand copied from the damaged template. (The vacancy left in the normal sister template strand is replaced by new synthesis and ligation.) Thus, between the two daughter duplexes there will be three strands carrying the original genetic information, and a flawed fourth. This flaw might be removed by a supervening repair mechanism after the replication or by simple dilution during subsequent replications, invoking the same kind of rectification by sister-strand exchange.

Error-prone dimer bypass If DNA polymerase is stalled too long without remedy, it might ignore the blockade (with the help of a protein synthesized in response to DNA damage by UV) and incorporate bases without regard to the fault in the template and continue its replicative activity. This bypass is the chief contributor to UV-induced mutations because it waives the editorial process.

THE SOS RESPONSE

DNA damage by UV light can induce the synthesis of a number of proteins to cope with the emergency. The genes with "SOS functions" are turned on when the recA protein, normally participating in homologous recombination reactions, is activated by effectors (probably breakdown products of damaged DNA) to become a protease. This causes the cleavage of the lexA protein, which is the repressor of the recA gene, as well as a number of other genes. These include those encoding the complex uvr endonuclease, the protein that promotes DNA replication across an unrepaired site on the template, the photoreactivation enzyme, and several other proteins involved in DNA repair. These multiple effects account for increased genetic recombination and resistance to further DNA damage.

Repairs of UV damage are very important to eukaryotic organisms as well as bacteria. There are humans with a rare skin disease known as xeroderma pigmentosum. Fibroblasts cultured from these patients are defective in the excision-repair system for removing pyrimidine dimers. These patients are unusually sensitive to sunlight, and exposure subjects them to great risks of developing cancer. Defects in at least seven different genes have been linked to this clinical condition.

Questions

6.1. What kind of mutations might or should be "silent"?

6.2. A wild-type gene specifies, in a particular position of an mRNA the condon UCG, for serine. Among possible single-base changes of this codon are: UUG (leucine), UCA (serine), UAG (nonsense), and UGG (tryptophan). (a) What is the likely phenotypic consequence for the mutation to UUG, UAG, and UGG? (b) Which of the mutations with likely phenotypic consequences can be stimulated by 2-aminopurine, hydroxylamine, or acridine orange?

6.3. How can a suppressor mutation in tRNA for tyrosine be tolerated?

6.4. Given a gene mutated by insertion of a single base pair into the beginning region of its sequence, would any of the following restore functionality?

(a) Single base pair substitution
(b) Single base pair addition
(c) Nonsense mutation
(d) Deletion removing ten base pairs
(e) Nonsense suppressor
(f) Deletion of 2 base pairs.

6.5. Four his^- mutants of E. coli (induced by ultraviolet light) were isolated from culture after enrichment with penicillin. The characteristics of these four mutants are listed below:

Mutant 1: Does not revert even after mutagenesis with nitrosoguanidine, ultraviolet light, acridine, or mustard gas.
Mutant 2: Reverts at a frequency of 10^{-6}. This reversion is stimulated by hydroxylamine.
Mutant 3: Reverts at a frequency of 10^{-6}. In one rare "revertant," the His^+ phenotype was found to be due to a second mutation which mapped in the gene coding for a tyrosine tRNA molecule.
Mutant 4: Reverts at a low frequency of 10^{-8}. This reversion was stimulated by acridine but not by nitrosoguanidine or hydroxylamine.

For each of the mutants, select the one most likely explanation from the following: transition, transversion, nonsense, missense, deletion, insertion, frameshift.

6.6. How can a *structural* gene for an enzyme — as opposed to a regulatory gene — be identified?

6.7. Three mutants unable to grow on lactose have been isolated. They all grow normally on glucose. Below are the characteristics of these mutants:

(a) The first mutant has no detectable β-galactosidase activity but has normal transport activity. Using antibodies, one can show that a β-galactosidase protein with a normal molecular weight is produced. The mutation reverts, and this reversion is stimulated by nitrosoguanidine but not by acridine.

(b) The second mutant has no detectable β-galactosidase or lactose transport activity. Specific antibodies against β-galactosidase precipitate no cross reacting proteins in cell extracts. No revertants are found.

(c) The third mutant shares the same phenotype with the first mutant, except the inactive β-galactosidase protein produced in this case is smaller than wild-type β-galactosidase.

What kind of mutation can be assigned to each mutant and in what gene did each mutation occur?

6.8. (a) A point mutation in the gene for adenyl cyclase (cya) results in the absence of $3',5'$-cyclic AMP in E. coli. Such cya mutants are no longer able to catabolize lactose, maltose, arabinose, and a number of other carbon sources. Explain. (b) Revertants of the cya mutant are isolated which can catabolize lactose, but not maltose or arabinose. Give two explanations for these revertants.

6.9. A mutant that fails to utilize L-arabinose has been isolated. Genetic tests show that the defect is a nonsense mutation. An assay shows that even cells grown in the presence of arabinose contain nondetectable activities of all three enzymes of the pathway. The loss of all three enzymes by a single point mutation immediately suggests that the mutation is in the regulatory gene. What other explanations would be possible and how would one distinguish the possibilities?

6.10. (a) What would be the effect of a nonsense mutation in the leader polypeptide of the histidine operon? (b) Would such a mutation be dominant?

6.11. How does one determine whether two genes belong to a single operon?

6.12. Insertion mutation by certain IS sequences were originally recognized through their very strong polar effects, for example, insertion of IS1 anywhere in gene A being totally polar on genes B and C (no detectable gene products). However, not all insertion sequences give polarity. Insertion of IS2 in one direction is polar, but insertion at the same site in the opposite direction is not. Speculate on what may be happening.

6.13. Describe the likely consequences of mutational loss of the transcription stop signal(s) for a bacterial gene. What would be the effect on the cell of a mutation that disrupts ρ protein?

Further Reading

Ames, B. N. 1979. Identifying environmental chemicals causing mutations and cancer. *Science* 204:587–593.

Demple, B., and P. Karran. 1983. Death of an enzyme: suicide repair of DNA. *Trends in Biochemical Sciences* 8:137–139.

Glass, R. E. 1982. *Gene function: E. coli and its heritable elements.* London: Croom Helm.

Hanawalt, P. C., P. K. Cooper, A. K. Ganesan, and C. A. Smith. 1979. DNA repair in bacteria and mammalian cells. *Annual Review of Biochemistry* 48:783–836.

Haseltine, W. A. 1983. Ultraviolet light repair and mutagenesis revisited. *Cell* 33:13–17.

Kenyon, C. J. 1983. The bacterial response to DNA damage. *Trends in Biochemical Sciences* 8:84–87.

Kornberg, A. 1980. *DNA Replication.* San Francisco: Freeman.

Lindahl, T. 1982. DNA repair enzymes. *Annual Review of Biochemistry* 51:61–87.

Loeb, L. A., and T. A. Kunkel. 1982. Fidelity of DNA synthesis. *Annual Review of Biochemistry* 51:429–457.

Nossal, N. G. 1983. Prokaryotic DNA replication systems. *Annual Review of Biochemistry* 53:581–615.

Roth, J. 1974. Frame shift mutants. *Annual Review of Genetics* 8:319–346.

Singer, B., and J. T. Kusmierek. 1982. Chemical mutagenesis. *Annual Review of Biochemistry* 51:655–693.

Tomizawa, J.-I., and G. Selzer. 1979. Initiation of DNA synthesis in *Escherichia coli. Annual Review of Biochemistry* 48:999–1034.

7

Genetic Exchange between Bacteria

BACTERIA ARE haploid organisms—each cell has only a single chromosome that has been inherited from a single parent. Nevertheless, sexual matings do occasionally occur, involving the transfer of genes from one bacterial cell to another. These transfers are not necessary for the generation of new offspring; but they play important roles in the maintenance of genetic integrity and the evolution of a bacterial population.

Gene transfer—the passage of DNA from one bacterial cell to another—is currently known to occur in three ways: through transformation, conjugation, and transduction. Transduction will be discussed in Chapter 10.

Transformation

Bacterial transformation, the first mechanism of gene transfer to be discovered, was demonstrated during the course of a study of *Streptococcus pneumoniae* (the pneumococcus) by F. Griffith (1928). This organism is the cause of classical bacterial pneumonia. The colonies of virulent strains have a characteristic smooth (S) morphology owing to the presence of a capsule. Colonies of strains lacking a capsule have a rough (R) appearance and are not pathogenic. Laboratory culturing of virulent strains often led to loss of virulence, which correlated with a shift from S to R morphology. R forms are probably mutants of S forms which have lost the ability to make capsules,

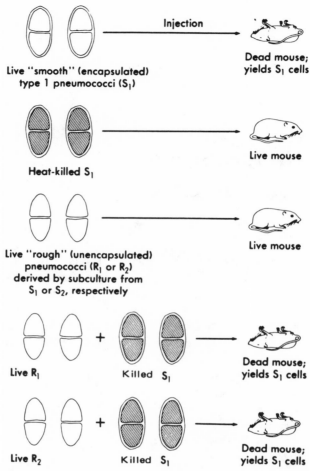

Fig. 7.1 The Griffith experiment. (From B. D. Davis, R. Dulbecco, H. N. Eisen, and H. S. Ginsberg. 1980. *Microbiology*. 3rd ed. New York: Harper and Row, p. 140.)

thereby attaining some selective advantage (such as faster growth) in culture.

Griffith showed that whereas the injection of R cells alone did not kill mice, the injection of live R cells and *killed* S cells was lethal (Fig. 7.1). Furthermore, bacteria recovered from the dead mice had somehow acquired the capsular type of the S strain. If the R cells were injected with killed S_1 cells, the cells recovered were of the S_1 type.

On the other hand, if killed S_2 cells were used, the cells recovered were of the S_2 type. In 1944 Avery, MacLeod, and McCarty showed that the experiment could be done in a test tube, and that the material being contributed by the dead cells was DNA: specifically, the genetic information for capsule type.

In general molecular terms, the Griffith–Avery experiment can be described as follows: Intact recipient cells either lack the gene(s) for capsule formation or carry a mutation in the gene. Capsule-producing cells killed by heat release DNA fragments large enough to contain ten or more genes. Transformation occurs when the recipient cells absorb this DNA by a special mechanism for DNA uptake. When this mechanism is operative in the cells, they are said to be **competent.** (Most bacteria are not competent for DNA uptake, and until recently transformation was only known for a few species in the genera Streptococcus, Hemophilus, and Bacillus.) However, there are now techniques for making other bacteria artificially competent, for example, by exposing them to a high concentration of Ca^{++}, which causes transient envelope changes.

Figure 7.2 shows the genetic state of a cell which has taken up a DNA fragment containing the gene(s) for capsule formation. The

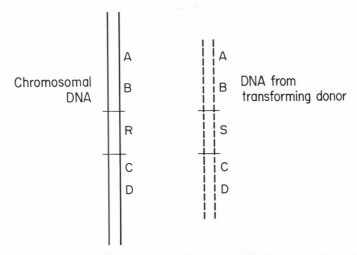

Fig. 7.2 The genetic makeup of a merodiploid cell during a transitional stage in transformation. The defective allele for the cell surface property is denoted as R (for rough colony), and the allele(s) that endows the capsule structure is denoted as S (for smooth colony).

genetic region shown is present in the cell in two copies, making it a partial diploid (**merodiploid**) for those particular genes. Other cells in the same culture will have taken up other fragments of DNA and will be merodiploid for other genes. Such a merodiploid state is transient —whereas the chromosome can replicate, the fragment cannot. In normal growth the fragment would be diluted out of the population and eventually degraded. However, as shown in Figure 7.2, the transforming fragment is exactly homologous in DNA sequence to the corresponding stretch of chromosome, except for the single mutation, for which the merodiploid is heterozygous (R/S)—that is, it contains two alternative versions of a gene at the same chromosomal locus. Each particular version of the gene is called an **allele.** In this situation a process called **homology-dependent recombination** can occur (Fig. 7.3). A **crossover** on each side of the region of interest, involving breaking and joining of the DNA molecules, can place the S allele in the chromosomal DNA. Crossover is an enzymatic process. (In *E. coli*, homology-mediated recombination is governed by the *recA* gene; *recA* mutants cannot carry out such recombination but can still grow.)

The result of the two crossovers in the pneumococcus is to endow the host cell with virulence. But even in the pneumococcus, transformation is a rare event under natural conditions. The degree of competence for DNA uptake is dependent on growth conditions. Furthermore, in the environment there are DNA nucleases (DNases)

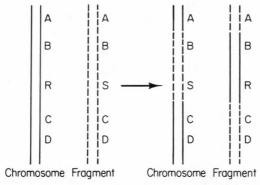

Fig. 7.3 Homologous crossing-over of the two sets of genes in Figure 7.2, resulting in genetic recombination.

which can destroy DNA. Even if surviving fragments are taken up by the cells, integration of the exogenous genes into the chromosome may not occur. The transformation of a particular genetic trait is even rarer. Therefore selective pressure must be present to reveal the event. In Griffith's original experiment, selection was based on the ability of a single virulent pneumococcus cell (or at most a few) to give rise to a progeny population large enough to kill the mouse. Generally, of course, selections are simpler, as discussed in Chapter 5. Recombinant selection is a fundamental step in genetic analysis.

The homology-dependent recombination discussed above is known in all cells and is often called **general recombination.** Other forms of recombination exist. One, called site-specific recombination, will be discussed in Chapters 8 and 11.

Conjugation

Joshua Lederberg discovered conjugation in *E. coli* by deliberate search. He reasoned that if recombination between bacteria occurred, it would be rare. He therefore prepared two mutant strains (which happened to be of opposite mating type), each differing in two characteristics: respectively, leu^-arg^- and his^-pro^-. Neither would grow on minimal medium without the two amino acids, and no double revertants would arise, since the frequency of both mutations reverting is the product of the reversion frequency of each ($10^{-6} \times 10^{-6}$). He found that when the particular pair of such double mutants were spread together in large numbers on the minimal plate, a few colonies grew (about 1 in 10^7 cells). The process had the following special characteristics:

(1) In contrast to transformation, intact donor *and* recipient cells were needed. Boiled cells or spent culture medium would not replace either strain. In fact, cell contact was required.

(2) *E. coli* strains could be classified by such experiments into two mating types: donors, called F^+ (for fertility), and recipients, called F^-. Recombinants were found in $F^+ \times F^-$ crosses, even more rarely in $F^+ \times F^+$ crosses, and never in $F^- \times F^-$ crosses.

(3) If several mutations were present in each strain, but only certain traits were selected, the recombinant clones generally

had the unselected characteristics of the F^- parent. That is, it appeared that the contributions of the two parental strains to the recombinant were unequal.

(4) In an $F^+ \times F^-$ cross the fertility characteristic itself acted as if it were transmissible at a high frequency even though chromosomal genes such as *his* and *leu* were found to recombine rarely.

THE F PLASMID

It was found that the fertility character was associated with a small extra chromosome, the F plasmid (see Chapter 8). Its size is about 2% of the bacterial chromosome in *E. coli*. F^+ cells carry this plasmid as a separate self-replicating entity. The F plasmid replicates independently of the bacterial chromosome and at a rate such that there are one or two copies per bacterial chromosome. When cells divide, an occasional daughter cell will accidentally fail to acquire a copy of the F plasmid (less than 10^{-3}). These cells are then F^-.

The F plasmid has two unusual properties: it is self-transmissible, and it can mediate the transmission of chromosomal genes from the host cell to the recipient. When an F^+ and F^- culture are mixed, the F plasmid spreads to almost all the cells, even if F^+ cells were a minority originally. F, thus acting as an infectious agent, converts F^- cells to F^+.

A set of genes on the F plasmid determines its ability to transfer itself to an F^- cell. These genes code for about thirty proteins involved in the complex transfer process and are highly expressed in comparison with similar genes carried by other plasmids (see Chapter 8). Some of these proteins determine the **F pilus.** The F pilus probably makes contact with a receptor on F^- cells, which may be a particular component of the outer membrane. This causes the two cells to form a conjugational bridge. Next, in a step sometimes called **sex-factor mobilization,** a special process of directional DNA synthesis drives one strand of the F plasmid into the recipient cell, where it is replicated, giving a second complete F plasmid. The remaining strand in the original cell is also replicated, so both donor and recipient cell inherit a daughter of the original F plasmid. During the process the replication fork remains attached to the conjugal bridge. The DNA intermediate resembles the letter σ, hence the name σ replication (Fig. 7.4).

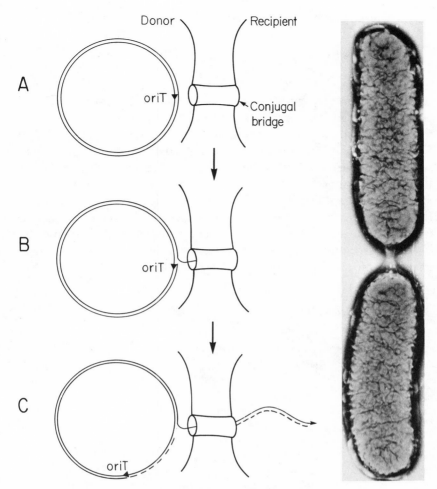

Fig. 7.4 Diagrammatic representation of the conjugal steps involved in F-plasmid transfer. *(A) Contact and pair formation.* The initial contact of cells is probably the result of random collision. Once in contact, the proteinaceous F pilus plays a role in the formation of a specific conjugation tube between the donor and the recipient cells. *(B) Sex-factor mobilization and transfer.* The mobilization and transfer of the F plasmid from a donor to a recipient involves a specialized form of DNA replication. An early mobilization step is an enzymatic "nick" of a circular F plasmid. This exposes a $5'$-phosphate end of a *single* DNA strand, which then enters the recipient. *(C) Establishment.* The entry of the single strand of F plasmid DNA is accompanied by the synthesis of a complementary strand within the recipient cell. Once synthesis and replication have been completed, the recipient, in turn, becomes a genetic vehicle for sex-factor transmission. The chromosomes of the mating cells are not shown. *(At right)* Electron micrograph of a conjugating pair of *E. coli.* (By R. N. Goldstein.)

HFR DONORS

The model described above accounts for transmission of the F plasmid itself from one cell to another, but not for the rare transfer of genes on the bacterial chromosome from F^+ to F^- cells. This ability is found in mutants of F^+ cells which transfer bacterial genes at high frequency. Such mutants are called **Hfr strains.** Hfr strains do not seem to contain the F plasmid in the usual independent form, for they do not convert F^- cultures at high frequency to F^+. However, Hfr strains do contain all the genes of the F plasmid itself. To explain these characteristics, Allan Campbell suggested that a cell is changed into an Hfr strain by a recombination between an F plasmid and the bacterial chromosome (Fig. 7.5) that integrates F into the bacterial chromosome as a linear stretch of DNA. Expression of the integrated

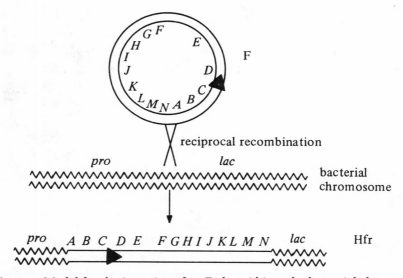

Fig. 7.5. Model for the insertion of an F plasmid into the bacterial chromosome. The circular F plasmid contains the hypothetical genes A to N; the origin of DNA transfer is depicted by an arrow between C and D. The jagged line represents a part of the much larger bacterial chromosome containing the bacterial genes pro and lac. After a pairing between homologous regions, a reciprocal crossover between F and the chromosome results in the linear insertion of F into the host genome to form an Hfr cell. The transfer of genes by the newly constituted chromosome follows the same scheme as depicted in Figure 7.4. (From S. Falkow. 1975. *Infectious Multiple Drug Resistance.* London: Pion, p. 13.)

F genes maintains the sex pili and the ability to transfer genes. However, the DNA that is transferred will now be that of the donor chromosome with its integrated F plasmid. Since the chromosome is so much larger than F, it takes almost two hours for the whole chromosome to be transferred from an Hfr cell to an F^- cell. The conjugation bridge of mating pairs is not very stable and often breaks, so partial chromosomal transfer is usual.

If the process of transfer is physically interrupted at different time intervals and the number of recombinants is determined, one finds that certain genes enter the recipient cell earlier than other genes (Fig. 7.6). In the example above, the F plasmid has integrated near *pro*, so some zygotes have received *pro* but not *thr-leu*; zygotes produced by longer mating have received both of these but not *arg*, and so on. Hfr genes which enter the recipient but fail to recombine with the resident chromosome by homologous crossover, as in the example discussed for pneumococcal transformation, will be lost by dilution or degradation.

The F plasmid carries several short DNA sequences that are homologous to sequences in the bacterial chromosome. These sequences are IS (insertion sequence) elements (see Chapter 8). Hfr formation can usually be explained by recombination between a plasmid IS and a homologous sequence on the chromosome. Thus, many different Hfr strains may be obtained, and each will transfer genes in a characteristic order. The order is consistent with a circular bacterial genome, the different Hfr strains having integrated the F genes at different points on the bacterial chromosome. At any given point of integration the F plasmid may be inserted clockwise or counterclockwise. The chromosomal gene first transferred into the recipient cell in one case would be the last gene to be transferred in the other case (Fig. 7.7).

Finally, the reason why transfer of the male (donor) property is so rare in Hfr $\times$ F^- matings is that a considerable portion of the DNA of the F plasmid is at the end of the chromosomal material being transferred. Unless the transfer is complete, recombinants arising from such crosses will not contain all the genes of F, and therefore will remain F^-.

The experiments with Hfr strains demonstrated the general order of genes on the *E. coli* chromosome and the fact that there is only a single chromosome. Hfr strains are used routinely for the mapping of genes (see below).

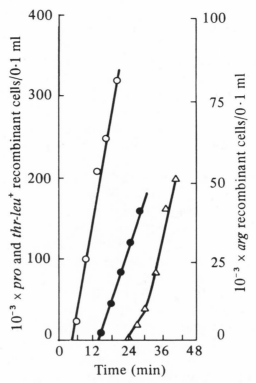

Fig. 7.6. Kinetics of gene transfer by *E. coli* Hfr strain. A mating was performed between an Hfr strain which was sensitive to streptomycin and an F^- strain which was resistant to streptomycin and required proline, threonine-leucine, and arginine. At timed intervals samples were removed and placed in a blender for 30 seconds. After interruption of the mating, samples were diluted and plated on media containing streptomycin but lacking one of the amino acids required by the recipients. The recombinant cells (open circles for pro^+, closed circles for thr^+ and leu^+, and triangles for arg^+) growing on these selective media are plotted as a function of time-sampling. (From S. Falkow. 1975. *Infectious Multiple Drug Resistance*. London: Pion, p. 11.)

F′ PLASMIDS

In the Campbell model for Hfr formation, a single crossover joins two circles. This process is reversible, as shown by the fact that it is usually possible to recover rare F^+ cells from an Hfr culture. This reverse process also involves genetic recombination. However, *aber-*

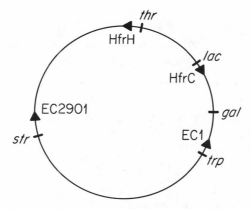

Fig. 7.7 Chromosomal gene transfer in *E. coli*. The circular chromosome is broken at the site of the origin of the transfer (arrows) and genes are transferred in linear order. Thus, for the HfrH strain, the order of genetic transfer would be *thr-lac-gal-trp-str*. For HfrC the order would be *lac-thr-str-trp-gal*. And so on for other strains.

rant recombination can also occur with crossover not in the original place (Fig. 7.8). Such an event may give a circular molecule of DNA carrying all of the F genes plus some of the bacterial genes that were near F on the bacterial chromosome. This aberrant F plasmid is called F′. **F′ plasmids** carrying blocks of genes from many different regions of the *E. coli* chromosome have been collected. Some carry blocks of genes as long as 20% of the total genome. F′ plasmids, like F, transfer themselves with high efficiency. The recipient of such a factor is merodiploid for that set of bacterial genes carried by the F′. Unlike the cases of transformation or conjugation between Hfr strains and F⁻, merodiploids of this type are not transient, for both the bacterial chromosome and the F′ genes can replicate. They are therefore permanent merodiploids and are useful for complementation tests (see below).

Genetic Mapping

In this and previous chapters we have referred to the location of genes relative to one another on the bacterial chromosome. The process of determining where genes are on chromosomes and how

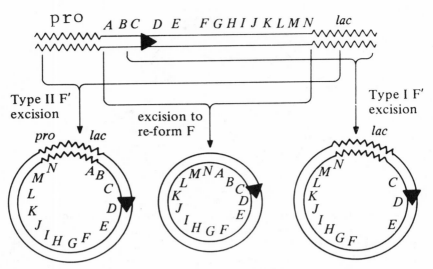

Fig. 7.8 Excision of F from an Hfr chromosome and the formation of different possible F′ plasmids. F inserted into a bacterial chromosome can, by excision, revert to the autonomous F$^+$ state. Occasionally, because of an aberrant excision process, a circular F′ plasmid carrying F genes and some bacterial genes is formed. In a type I excision, one cut is made within F and the other in the bacterial chromosome. This example leads to an F′ which has lost the F genes *AB* and has gained the bacterial gene *lac*. In a type II excision, the cuts are made within the bacterial sequences on *both* sides of F. In the example, this leads to an F′ carrying all the F genes and both the bacterial genes *pro* and *lac*. (From S. Falkow. 1975. *Infectious Multiple Drug Resistance*. London: Pion, p. 17.)

mutations are ordered within genes is called genetic mapping. There are several methods of analysis. The first, **mapping by interrupted mating,** was described in the section above (Figs. 7.6 and 7.7).

THE TWO-FACTOR CROSS

As the distance between two mutations increases, so does the frequency of recombination between them. This principle has been employed to order the location of nearby mutations. Suppose that a series of *lac*$^-$ *E. coli* mutants is isolated, and interrupted mating shows the mutations to lie all in the same region of the chromosome (at about 10 minutes). Though interrupted mating cannot discriminate between them, they could be mapped by recombinational frequencies in a series of pairwise crosses: HfrH $lac1$ str^s $\times$ F^- $lac2$ str^r, selecting lac^+ str^r. (On an agar plate containing lactose as the only carbon and energy source and the drug streptomycin, only the F^- exconjugants that received a portion of the intact lac gene to replace the $lac2$ mutant site by recombination will grow.) This might give the following results:

Mating pairs		Frequency of
Male	*Female*	lac recombinants
$lac1$ $\times$	$lac2$	1×10^{-4}
$lac1$ $\times$	$lac3$	0.3×10^{-4}
$lac1$ $\times$	$lac4$	0.5×10^{-4}
$lac2$ $\times$	$lac3$	0.5×10^{-4}
$lac2$ $\times$	$lac4$	0.4×10^{-4}
$lac3$ $\times$	$lac4$	0.1×10^{-4}

The set of data suggests the following order: $lac1$-$lac3$-$lac4$-$lac2$. Such a technique may be refined by performing the experiments reciprocally and co-selecting for an outside marker, such as pro (Fig. 7.9). In this example, if the order were $lac1$-$lac2$-pro^-, cross A would give many more recombinants than cross B, and vice versa.

THE THREE-FACTOR CROSS

Figure 7.10 depicts the cross of Figure 7.6 in a different way. The appearance of recombinant colonies depends on the formation of a partial heterozygote: to obtain an arg^+ recombinant at least the portion of chromosome between the origin and arg must have entered from the Hfr. At least two crossover events, one on each side of arg, giving an arg^+ str^r recombinant are also necessary. If the cross were done selecting only for arg^+ recombinants (on minimal medium with streptomycin and proline and threonine), the distribution of the other alleles — pro^+ or pro^-, thr^+ or thr^- — could be scored as independent markers. The four possible recombinant classes would occur in a frequency depending on the order of and distance between the genes, for example:

pro^+, thr^+ (arg^+): 40% pro^+, thr^- (arg^+): 5%
pro^-, thr^+ (arg^+): 30% pro^-, thr^- (arg^+): 25%

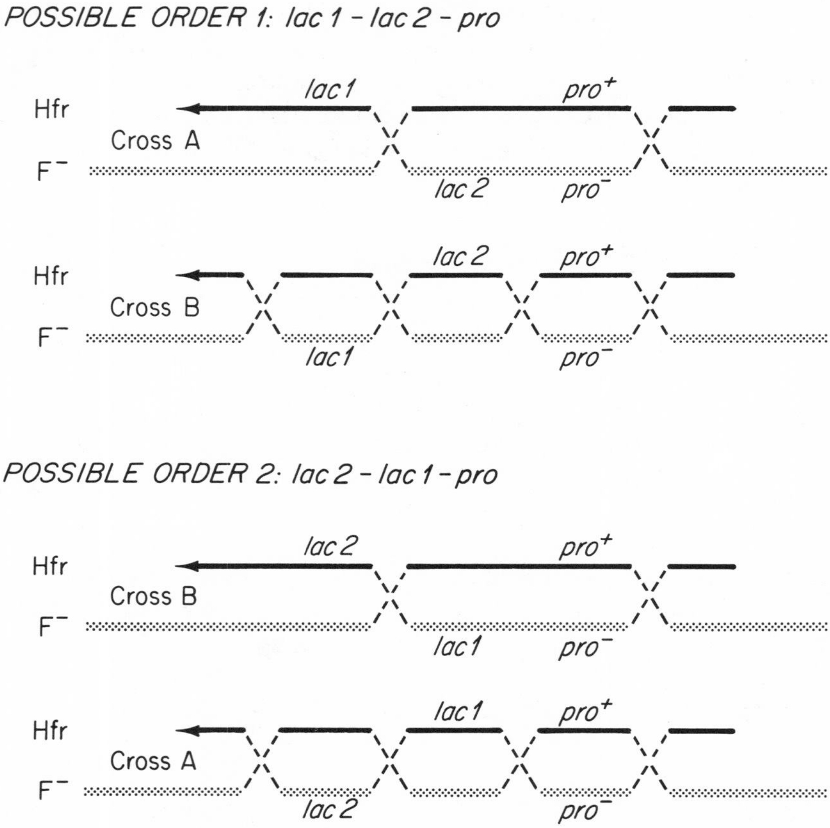

POSSIBLE ORDER 1: lac 1 – lac 2 – pro

POSSIBLE ORDER 2: lac 2 – lac 1 – pro

Fig. 7.9 A three-point genetic cross to determine the order of two mutations within the same gene. In all crosses, the Hfr strain carries pro^+ and the F^- strain carries pro^- and the recombinants are selected for the ability to grow on lactose without proline supplementation. If the topological order of mutations is $lac1$-$lac2$-pro^- (as shown in the top half of the figure), then fewer lac^+ and pro^+ recombinants will be found when the $lac1$ mutation is placed in the F^- strain and the $lac2$ mutation is placed in the Hfr (quadruple crossovers required) than in the reciprocal cross in which the $lac1$ mutation is placed in the Hfr strain and the $lac2$ mutation is placed in the F^- strain (double crossovers required). The opposite results will be obtained if the topological order of mutations is $lac2$-$lac1$-pro^- (as shown in the bottom half of the figure).

The rarest class would necessitate *two* pairs of crossover events. The data from this analysis of *thr* and *pro* as "unselected markers" thus give the same gene order (ori - pro - thr - arg) as the interrupted mating. Such recombinational analysis is a more precise technique of genetic mapping than interrupted mating.

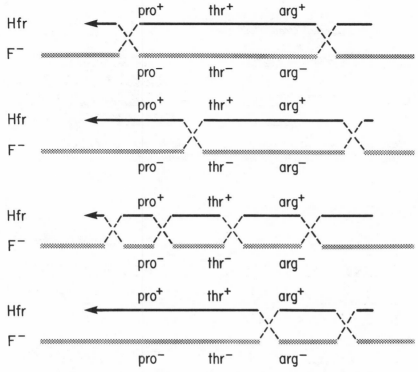

Fig. 7.10 A three-point genetic cross in conjugation. The donor Hfr strain carries the markers pro^+, thr^+, arg^+, and str^s (streptomycin-sensitive; marker not shown because it is far distal to the entering end of the Hfr chromosome indicated by the arrow). The recipient F^- strain carries the markers pro^-, thr^-, arg^-, and str^R (streptomycin-resistance). After mating, the cells were spread on agar containing streptomycin, proline, threonine, and glucose. Hfr cells are killed by the drug. F^- cells that have not acquired the arg^+ gene will not give rise to visible colonies, because the amino acid is not provided in the medium. Recombinants that have acquired the arg^+ gene may or may not have also acquired thr^+ and pro^+, depending where the crossovers (dashed lines) occurred. The least frequent class of recombinants would have the genotype arg^+, thr^-, and pro^+, since quadruple crossovers are required.

DELETION MAPPING

The most unambiguous way to map closely linked mutations is by using deletions (Fig. 7.11). This has the advantage of giving yes or no answers. For if the region deleted from one chromosome covers a particular mutational site on the other chromosome, no recombination can yield the wild-type gene again. Of course, such analysis

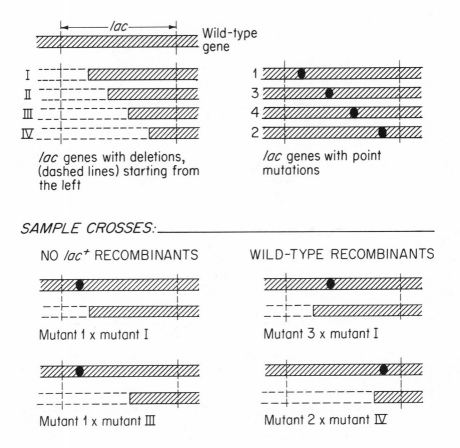

lac genes with deletions, (dashed lines) starting from the left

lac genes with point mutations

SAMPLE CROSSES:

NO *lac⁺* RECOMBINANTS

Mutant 1 x mutant I

Mutant 1 x mutant III

WILD-TYPE RECOMBINANTS

Mutant 3 x mutant I

Mutant 2 x mutant IV

Fig. 7.11 Deletion mapping of a *lac* gene. No recombinants will arise in crosses between a deletion mutant and a point mutant whose defect is in the region of that deletion. Mutant IV, with the largest deletion, will therefore fail to give wild-type recombinants when crossed with point mutants 1, 3, and 4. A complete pairwise cross should define both the position of the point mutations and the position and extent of the deletions.

requires a set of deletion mutations for the gene in question. A number of techniques exist for obtaining them.

PROBLEMS IN RECOMBINATIONAL ANALYSIS

The molecular mechanisms of homology-dependent recombination are not completely understood. Without this knowledge we cannot accurately predict how two different mutations will behave with respect to each other in a recombinational analysis. Recombinational mapping techniques are predicated on the assumption that homology-dependent recombination occurs in a strictly random fashion, that is, the chance of the event occurring between any two mutations on the chromosome is directly related to the distance between them: the greater the distance, the greater the recombinational frequency. Generally, this is true, but exceptions are common. For example, recombination "hot spots" have been discovered. If such a hot spot is located between two mutations, the recombination frequency between them will be abnormally high, suggesting that the mutations are farther apart than they actually are. For this reason two-factor mapping is rarely performed; three-factor or deletion mapping are the methods of choice.

Complementation Analysis

The recombination analyses described above provide a formal method for determining the location of mutations and genes. However, this approach alone cannot tell how many genes participate in determining a given phenotype, such as the ability to grow on lactose. For defining the units of genetic function, complementation analysis is a very useful tool. Complementation studies involve the examination of the phenotype of cells that carry two copies of a genetic region, each with a different mutation, for example, two different lac^- mutations.

COMPLEMENTATION WITH F′ PLASMID

In practice there are several ways by which *stable* merodiploids can be constructed. The easiest and most common method involves the use of plasmids that carry bacterial genes, such as F′*lac*. As a preparative step, the lac^- mutations 1, 2, 3, and 4 are introduced into F' lac by recombination. E. $coli$ strains with F's carrying the various versions of lac^- are then individually mated with the various F$^-$ lac^- mutant strains. The resulting stable merodiploids are then scored for their Lac phenotype, which might be as follows:

Merodiploid	Lac phenotype
$lac1/lac2$	Lac$^+$
$lac1/lac4$	Lac$^+$
$lac1/lac3$	Lac$^-$
$lac2/lac3$	Lac$^+$
$lac2/lac4$	Lac$^-$
$lac3/lac4$	Lac$^+$

These results show that the four lac mutations fall into two **complementation groups,** A and B. Generally, a complementation group defines a gene. Thus the data resolve the lactose system into two genes:

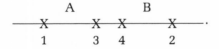

Mutations 1 and 3 are in the same complementation group (A); this says that the two mutations impair the same gene product, and the two defective products cannot provide normal function. Similarly, mutations 2 and 4 are in the other complementation group (B). The mutations in the different complementation groups (A and B) complement each other; in this case the deficiency of gene product A resulting from the mutation in one chromosome is compensated by the functional gene A on the other chromosome, and the reciprocal applies to gene product B. The rationale behind this concept is diagrammed in Figure 7.12.

It is very important to note that complementation does not involve a recombination event. Complementation between two lac mutations in merodiploid cells renders every cell competent to metabolize lactose. In contrast, recombination between two lac mutations occurs at much lower frequency, and only 1% or fewer of the diploid cells will be Lac$^+$. Thus, recombination and complementation are

COMPLEMENTATION OF FUNCTION

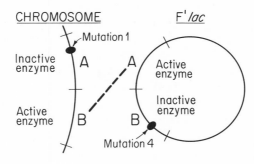

NO COMPLEMENTATION OF FUNCTION

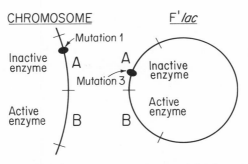

Fig. 7.12 Complementation between mutations 1 and 4, but not mutations 1 and 3, in a merodiploid state.

easily distinguishable by the frequency at which the Lac$^+$ phenotype appears.

PROBLEMS IN COMPLEMENTATION ANALYSIS

Complementation analysis may be complicated in numerous ways. One difficulty is introduced when the problem of **dominance** and **recessiveness** of various mutations has to be dealt with. In the examples given above, the complementation test was used for analyzing recessive mutations. If a mutation is dominant, then the presence of the wild-type gene will not restore the original phenotype (see the following section on regulatory mechanisms). A mutation can also be **co-dominant** with the wild-type allele. An example would be a mutant producing an enzyme with altered substrate specificity. If the wild-type allele is also present, then the collective properties of the two enzymes will be manifested.

Restoration of the original phenotype need not imply the introduction of a wild-type allele which is dominant over the mutant allele. An example of **nonallelic dominance** is the nonsense suppression resulting from an alteration of a tRNA species (discussed in the previous chapter). Through the action of the new tRNA, an amino acid is inserted at a termination codon resulting from a mutation in a structural gene. In a supressor strain, there are populations of both suppressor and wild-type tRNAs. Thus, in a strain which is a partial diploid of sup^+ (mutated tRNA)/sup^- (wild-type tRNA), the nonsense mutation in the structural gene will not be phenotypically manifested. The sup^+ allele is said to be **epistatic** with respect to the nonsense mutation, but co-dominant with respect to the sup^- allele.

A final caveat should be made about the concept that one complementation group represents a single gene. First, numerous examples exist where a gene codes for a protein with more than one enzymatic or biological activity. Second, intragenic complementation is relatively common, if the protein product is an oligomer. Subunit mixing can often restore some enzymatic activity, as illustrated in Figure 7.13.

COMPLEMENTATION ANALYSIS OF REGULATORY MECHANISMS

The original model of lac expression was partly made on the basis of complementation analysis of the various regulatory mutations (see Chapter 5 and Table 7.1). Most merodiploids carrying both the wild-type (inducible, $lacR^+$) and mutant (constitutive, $lacR^-$) alleles of the $lacR$ gene are inducible, not constitutive. Apparently, enough normal repressor is made from the $lacR^+$ allele to bind to both copies of the lac operator in these diploids.

For the mutation $lacR^s$, the super-repressor cannot be removed from the operator by the inducer. Therefore, in the diploid $lacR^+$/$lacR^s$, all operators are tied up by the mutant repressor molecules, and the phenotype of these diploids is Lac$^-$.

O^c, or operator constitutive, mutations behave differently from the $lacR^-$ mutations in complementation analysis. When the cell carries two lac operons, one with a normal O^+ operator and the other with an

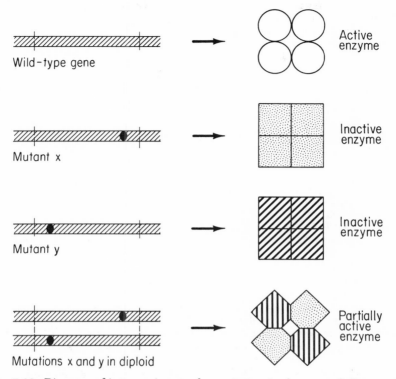

Wild-type gene — Active enzyme

Mutant x — Inactive enzyme

Mutant y — Inactive enzyme

Mutations x and y in diploid — Partially active enzyme

Fig. 7.13 Diagram of intragenic complementation in the case of oligomeric enzymes. Two abnormal protein subunits encoded by mutant allelic genes may by chance form a complex with mutual compensation, resulting in a functional product.

O^c operator, only the *lac* structural genes linked to the O^c operator are expressed constitutively. The O^c mutation affects only those genes of the mutated operon (said to be in "cis") and not those in a homologous operon of another chromosome, or even in another part of the same chromosome (said to be in "trans"). This kind of mutation is therefore **cis-dominant.** By the same token, the *lacR*^{s} mutation is **trans-dominant.**

Structural genes such as *lacR* or *lacZ*, which code for proteins, are fundamentally different from control sites such as *lacO* or *lacP* (itself containing two subdomains, the CAP site and the RNA polymerase-binding site; see Fig. 5.13), which merely serve to receive signals or act as binding sites. The operator is not a gene; neither is the promoter. Mutations in control sites necessarily only act in *cis*.

Table 7.1. Complementation analysis of regulatory mutations of the lactose system in *E. coli*.

Genetic constitution of haploids or merodiploids	Units of β-galactosidase/ mg cell protein	
	Without inducer	With inducer
$R^+ O^+ Z^+$	1	1,000
$R^+ O^+ Z^+ / R^+ O^+ Z^+$	2	2,000
$R^- O^+ Z^+$	1,000	1,000
$R^- O^+ Z^+ / R^+ O^+ Z^+$	2	2,000
$R^+ O^c Z^+$	500	1,000
$R^+ O^c Z^+ / R^+ O^+ Z^+$	500	2,000
$R^+ O^c Z^-$	0	0
$R^+ O^+ Z^+ / R^+ O^c Z^-$	1	1,000
$R^s O^+ Z^+$	1	1
$R^s O^+ Z^+ / R^+ O^+ Z^+$	2	2

R^+ wild-type repressor.
R^- repressor constitutive (the protein is either absent or acts as though it has lost the affinity for the operator site).
R^s altered repressor acting as though it no longer interacts with the inducer; consequently the expression of the *lac* operon is "superrepressed."
O^+ wild-type operator.
O^c operator constitutive. (The DNA site is defective in binding repressor.)
Z^+ wild-type structural gene for β-galactosidase.
Z^- mutant structural gene for β-galactosidase.
$/$ separates the description of the two genotypes in a merodiploid cell.

Complementation analysis does not always give the straightforward results seen in Table 7.1. Complications can arise from the fact that regulatory gene products are often oligomeric proteins, so that mixed oligomers with unpredictable properties can be formed in diploid strains. Thus, there exist *lacR*⁻ mutations that give the constitutive phenotype even in *lacR*⁻/*lacR*⁺ diploids. These complications can be avoided if deletions, frame shifts, or nonsense mutations, rather than missense mutations, are used for the analysis.

Many aspects of this model for expression of the *lac* operon have been verified through *in vitro* studies. Such studies included the use of *lac* DNA to direct cell-free synthesis of β-galactosidase, the isolation and characterization of the repressor, and the determination of the alterations in the DNA sequence caused by mutations in regulatory regions.

COMPLEMENTATION AND DOMINANCE IN DIPLOID ORGANISMS

In diploid organisms, the contribution of each allelic gene pair determines the overall phenotype. In humans, many genetic diseases are recessive, and thus are manifested only in the homozygous state. Complementation by the normal allele corrects the partial defect in the heterozygote. However, some mutations are manifested even in the heterozygous state and are therefore said to be dominant. Examples of recessive and dominant genetic defects are those seen in sickle cell anemia and Huntington's chorea, respectively.

Questions

7.1. Suppose that in transformation or conjugation the entering DNA did *not* recombine with the recipient chromosome. (a) Would the genes in the piece of DNA function? For example, would the merodiploid cell of Figure 7.2 produce capsule? (b) Would such a merodiploid grow to give a smooth colony?

7.2. As isolated from nature, most *E. coli* do not contain the F plasmid, but some do. (a) How might this fact have been determined? (b) If the F plasmid caused F^- cells to become F^+, why are not all *E. coli* in nature F^+?

7.3. A wild-type phenotype can be generated by either recombination or complementation. Explain the molecular basis for each and its utility in genetic analysis.

7.4. (a) Describe a method to select for mutants that are defective in an essential gene of *E. coli*, using RNA polymerase as an example. (b) RNA polymerase is made of different subunits. How could one show whether or not the mutants obtained in (a) affected different subunits?

7.5. Mutants of *E. coli* requiring tryptophan were analyzed and gave the following recombination frequencies:

Hfr × F$^-$	% Trp$^+$ recombinants
$trp^-1 \times trp^-2$	0.05
$trp^-2 \times trp^-3$	0.80
$trp^-3 \times trp^-4$	1.50
$trp^-4 \times trp^-1$	1.00
$trp^-4 \times trp^-2$	0.90
$trp^-1 \times trp^-3$	0.70

All the mutations could revert, and none were frame-shift or nonsense mutations. A series of complementation experiments with F'trp plasmids yielded the following data:

(1) trp^-1 could not complement trp^-2
(2) trp^-1 or trp^-2 could complement trp^-3
(3) trp^-4 could not complement any known trp^- mutation
(4) trp^-4/trp^+ diploid does not require tryptophan for growth.

(a) From the recombination data, determine the order of mutations on the genetic map. Draw the map, giving the relative distances between the mutation sites. (b) How many genes are defined by the set of mutations trp^-1, trp^-2, and trp^-3? (c) What type of mutation is trp^-4? (d) Another trp^- mutation on the chromosome was not complemented by an F'-trp^+ plasmid. What kind of mutation might this be?

7.6. In the genetic map in Figure Q7.6 are shown five deletion mutations. Their spans are indicated by solid bars (for example, deletion 1 is missing most of gene A). A strain bearing point mutation X is crossed with each of the 5 deletion strains; wild-type recombinants arise *only* from mating with deletions strains 1 and 5. In which gene(s) might point X mutation be located? Give the reason.

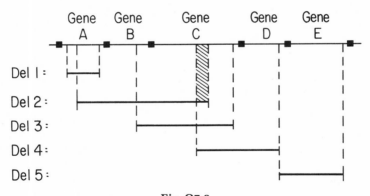

Fig. Q7.6

7.7. How can one distinguish true reversion from suppression?

7.8. Mutants lacking major cell wall components (such as capsule or lipopolysaccharide) often given a different colony appearance, "rough" instead of "smooth." Rough mutants usually are cleared by phagocytosis in an infected host even if there has been no previous exposure to them. In the case of the smooth parental strains, exposure of the host immune system to the antigen is necessary before the organisms are cleared. Explain.

7.9. For the following merodiploid $E.$ $coli$ strains, indicate if the expression of $lacZ^+$ gene is inducible, constitutive, or noninducible:

(a) $F'\ R^+\ O^+\ Z^+/R^s\ O^+\ Z^-$

(b) $F'\ R^+\ O^c\ Z^+/R^+\ O^+\ Z^-$

(c) $F'\ R^-\ O^+\ Z^-/R^+\ O^+\ Z^+$

(d) $F'\ R^s\ O^+\ Z^-/R^-\ O^c\ Z^+$.

Superscripts are defined as follows: "+" = wild-type allele; "c" = constitutive; "s" = super-repressor; "−" = function destroying mutation.

Further Reading

Achtman, M., and R. Skurray. 1977. A redefinition of the mating phenomenon in bacteria. In $Microbial\ Interaction\ (Receptors\ and\ Recognition)$, ser. B, vol. 3, ed. J. L. Reissig. London: Chapman Hall, pp. 233–279.

Bachmann, B. J. 1983. Linkage map of $Escherichia\ coli$ K-12, Edition 7. $Microbiological\ Reviews$ 47:180–230.

Birge, E. A. 1981. $Bacterial\ and\ Bacteriophage\ Genetics:\ An\ Introduction.$ New York: Springer-Verlag.

Dressler, D., and H. Potter. 1982. Molecular mechanisms in genetic recombination. $Annual\ Review\ of\ Biochemistry$ 51:727–761.

Goodgal, S. H. 1982. DNA uptake in $Haemophilus$ transformation. $Annual\ Review\ of\ Genetics$ 16:169–192.

Radding, C. M. 1981. Recombination activities of $E.\ coli$ RecA protein. $Cell$ 25:3–4.

——1982. Homologous pairing and strand exchange in genetic recombination. $Annual\ Review\ of\ Genetics$ 16:405–437.

Sanderson, K. E., and J. R. Roth. 1983. Linkage map of $Salmonella\ typhimurium$, Edition 6. $Microbiological\ Reviews$ 47:410–453.

Smith, H. O., D. B. Danner, and R. A. Deich. 1981. Genetic transformation. $Annual\ Review\ of\ Biochemistry$ 50:41–68.

Stahl, F. W. 1979. Special sites in generalized recombination. $Annual\ Review\ of\ Genetics$ 13:7–24.

——1979. $Genetic\ Recombination:\ Thinking\ about\ It\ in\ Phage\ and\ Fungi.$ San Francisco: Freeman.

Willets, N., and R. Skurray. 1980. The conjugation system of F-like plasmids. $Annual\ Review\ of\ Genetics$ 14:41–76.

8

Plasmids

PLASMIDS are circular, double-stranded extrachromosomal DNA molecules that contain specialized genes and have the ability to be replicated in the bacterial cell. Frequently they can move from one bacterial cell to another and may even be exchanged between cells of different species. These DNA molecules coordinate their replication with cell division. Often associated with plasmids are transposons and insertion sequences: recombination units that provide plasmids with unusual genetic diversity.

Plasmid-Borne Traits

The F plasmid, which has the ability to transfer genes from one bacterial chromosome to another (see Chapter 7), is not the only plasmid with the ability to change the characteristics of a recipient cell. Plasmids conferring resistance to antibiotics were discovered more than twenty years ago in Japan through the analysis of an epidemic of bacillary dysentery. Strains resistant to the four antibiotics then in wide use—sulfonamides, chloramphenicol, streptomycin, and tetracycline—became common. This pattern of multiple drug resistance could not be obtained in the laboratory by mutation of sensitive strains. Moreover, the property was found to arise by conjugational transfer from certain multiply resistant *E. coli*. Eventually, it was discovered that the resistance determinants were carried by plasmids. The term **R factors** is used widely to describe these genetic units, because the multiple resistance factors were not proven to be on plasmids until some time after their discovery.

R plasmids that code for the inactivation of deleterious agents other than antibiotics are also known. Ions of many heavy metals such as silver, mercury, and lead are toxic to bacteria. There exist R plasmids that confer resistance to each of these. For example, bacteria that grow in waters contaminated with photochemical effluents often carry a plasmid-coded enzyme that reduces silver ions to metalic silver. One of the goals of biotechnology is to harness such genes for the economical extraction of valuable metals from spent wastes or low-yielding ores.

Although R plasmids are the most conspicuous of the plasmids, they are certainly not the only ones of importance. In human disease, an extremely important class of plasmids are those carrying **virulence factors.** A well-studied example is enteritis caused by certain strains of E. coli that have two atypical characteristics: synthesis of adherence pili that allow the cells to colonize the upper intestine, and the ability to excrete a protein toxin (toxinogenic) that damages intestinal tissue. Genes for these two traits are sometimes borne on different plasmids.

Pseudomonads have the ability to digest some highly complex organic compounds, such as camphor, asphalt, high octane gasoline, and so on. Genes for the metabolic pathways permitting oxidation of these compounds to intermediates in central pathways are carried on plasmids. Much effort is being expended toward utilizing these plasmids to clean up oil spills and other organic wastes.

Another class of plasmids, the **bacteriocinogens,** instigate fratricidal warfare by causing the host to excrete proteins that kill conspecific or closely related bacteria not carrying the plasmid. At the same time these plasmids afford toxin immunity to their host cell. The lethal proteins made by E. coli are called colicins; the counterparts made by Pseudomonas pyocyaneas are called pyocins, and so on. Against a given bacterial species there are lethal proteins with different modes of action. For example, colicin E1 damages the cytoplasmic membrane, E2 attacks DNA, and E3 stops protein synthesis by cleaving off ribosomal subunits.

In general plasmid-borne traits provide survival value to the bacterial host under special conditions that may only exist intermittently (such as the presence of rare carbon sources or antibiotics). Since such conditions arise infrequently, we can see simply on the basis of the cost of maintaining extra DNA why these genes are not more permanently established within an entire species. A plasmid can be maintained in a subpopulation even without selective pressure, because of their tendency to spread by infection. Bacteriocinogens have a selective advantage in an even more specialized situation in that the competition would be with members of the same bacterial species which are vulnerable because they are not yet infected. If all the members of a local population were to enjoy the immunity acquired by infection, the plasmid would simply become a molecular parasite. Thus the bacteriocinogens are likely to persist only if a majority of the cells in the community are without them. An evolutionary view can also be applied to the toxinogenic *E. coli* plasmids that cause the infected subclone to subvert the symbiotic relationship of *E. coli* with its mammalian host.

PLASMID REPLICATION AND COPY NUMBERS WITHIN THE CELL

Different kinds of plasmids are in part characterized by their copy number (Table 8.1). For example, the sex plasmid F is present in 1 to 2.5 copies per copy of the *E. coli* chromosome, whereas some small plasmids may be present in 30 to 60 copies. The ratio of plasmid copy number to bacterial chromosome copy number is a more reliable indicator of the replicative property of the plasmid than plasmid number per cell because the number of genomes per cell varies with the rate of bacterial growth.

As with the bacterial chromosome, a round of plasmid replication begins by the initiation of DNA synthesis at a specific site called *ori* and proceeds in much the same way as bacterial replication (see Chapter 6). Control of the number of both bacterial chromosomes and plasmids is always exerted at the initiation stage. With large plasmids, replication is bidirectional, but some plasmids, especially smaller ones, have only a single active fork and are said to undergo *unidirectional* replication. The fork moves around the replicon until it returns to *ori*; the two double-stranded daughter molecules then separate.

In general, one of the prerequisites for a new round of DNA synthesis is the ability of the cells to carry out protein synthesis. In the case of the *E. coli* chromosome and many large plasmids, initiation of DNA synthesis can be prevented by the addition of chloramphenicol. Such plasmids are said to be under "stringent" replication

Table 8.1. Essential characteristics of some plasmids.

Plasmid	Size^a (kb)	Copy number^b	Relevant traits	Comments
Col E1	9	10-30	Colicin E1 production	Nontransmissible relaxed replication
R6K	40	10-40	Antibiotic resistance	Relaxed replication, tra^+
Ent (P307)	90		Heat labile and stable enterotoxins	Found in toxigenic strains of E. coli, tra^+
RP4	54	1-3	Amp, kan, neo, tet resistance genes	A broad host range plasmid, tra^+
F	95	1-3	Phage T7 resistance	E. coli sex factor
R6	98	1-3	Cml, kan, neo, str, sul, pur, tet resistance genes	Prototype R plasmid
P1	89	1-2	Phage P1 resistance	Prophage of P1, releases infectious particles

a. Size is in kilo (1,000) base pairs. For comparison, a gene that codes for an average sized protein of 40,000 daltons is about 1.2 kb.

b. Copy number is the ratio of plasmid copies to bacterial chromosome copies in the cell (measured in E. coli).

control. This implies that there is a specific cellular signal that permits replication. Many small plasmids, such as ColE1 (a plasmid specifying colicin E1), will continue to replicate in the presence of chloramphenicol. Such plasmids are said to be under "relaxed" replication control.

INCOMPATIBILITY AND COPY-NUMBER CONTROL

The following description of the mechanism whereby plasmids regulate their own replication applies to the small multicopy plasmid ColE1 but may not be entirely true for the larger R plasmids. In general, plasmids replicate independently of the bacterial chromosome, and at any given time only a single, randomly selected plasmid undergoes replication (Fig. 8.1). At the beginning of a cell cycle, it appears that as a consequence of its replication a plasmid produces a substance that diffuses to other plasmids of the same type, preventing

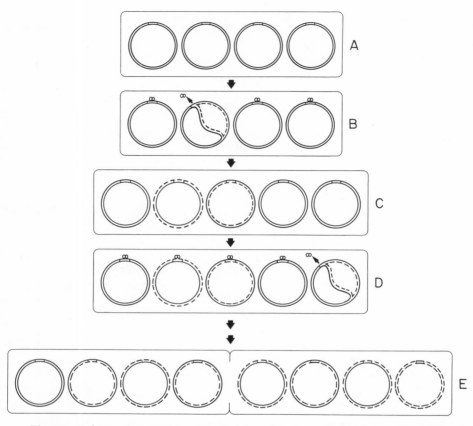

Fig. 8.1 Scheme for random sequential replication of a multicopy plasmid. The dimeric structure (in B and D) represents a diffusible inhibitor synthesized from an actively replicating molecule. In the case of plasmid ColEI, this inhibitor molecule is an RNA. Since many such molecules are unstable, the concentration in the cell depends not only on the rate of synthesis and dilution with cell division but also on the rate of degradation. Solid lines represent parental DNA. Initiation of replication is random. Therefore some plasmids are not copied in a given cycle of cell division, whereas others may be replicated more than once.

initiation at their *ori* site (an example of negative control). After the first round of replication is completed and the inhibitor is exhausted by decay, a second plasmid, again activated at random, undergoes a round of replication. Replication of this molecule will likewise prevent the other plasmids from initiating, and so on, until the cell divides again. Thus copy number is determined by the relative stability of the inhibitor and the time required to replicate a single plasmid.

The model predicts that if the plasmid is made larger by the insertion of additional DNA, then the plasmid copy number will decline, since rate of fork movement is constant. It is also expected that in any given round of cell division some plasmid molecules of a given type will not serve as a template at all, whereas some others might undergo two or more rounds of replication. At the end of the cell cycle each of the two daughters will inherit plasmid copies through the interaction of a specific partition site on the plasmid called *par* with a yet to be defined cellular structure that will be divided and inherited by the two daughter bacteria.

This model allows one to determine whether or not two plasmids have the same *ori* sequence. Two different types of plasmids, conferring different traits to the cell, may have identical sequences in the regions designated *ori* and *rep* (coding for proteins involved in DNA replication), even though their structures vary elsewhere. Owing to inhibition at the initiation stage, two types of plasmids with similar *ori* and *rep* will be unequally copied and hence unequally distributed into daughter cells. Sooner or later some new cells will have inherited copies of only one kind of plasmid. Two such plasmids that are unequally inherited are said to be **incompatible.** To test for incompatibility one only needs to introduce two different plasmids into a single cell by taking advantage of different plasmid traits (for example, by co-selecting for an antibiotic resistance conferred by one plasmid and a bacteriocin immunity conferred by the other) and then to determine whether or not the two traits partition equally to all of the daughter clones when selective pressure is removed. If the two plasmids can be maintained together indefinitely, they belong to different incompatibility groups (that is, their *ori* and *rep* regions are structurally different); if not, they are in the same group. Because the *ori* and *rep* regions are such fundamental features of the plasmid's structure, incompatibility has become a major criterion for classification of bacterial plasmids.

Gene Transfer from Cell to Cell

In addition to being self-replicating within a cell, many plasmids have a block of genes that enable the cell carrying the plasmid to act as a conjugal donor. These plasmids are classified as **self-transmissi-**

ble. However, such plasmids differ widely in their frequency of transmission, even between cells of the same species. The frequency can be as low as 10^{-4} for some R plasmids and as high as one for the F plasmid.

Figure 8.2 shows the essential genetic structure of an F plasmid. The block of *tra* (for transfer) genes code for at least four proteins that are needed for DNA synthesis during transfer; they also include genes for sex pili and often genes for **surface exclusion proteins** (located in both the outer and the inner membranes) that prevent other male cells from attaching to the host cell, and thus forestall futile conjugal transfer. The transfer process is like that described in Figure 7.4. The F plasmid is "primed" at *oriT* simply by a nick in one strand of the DNA duplex. However, the origin of genetic transfer is different from that of simple replication and division (*oriR*), and fork movement is undirectional. In fact, the order of gene transfer in an Hfr cross is determined by the direction of fork movement from *oriT*.

Most small plasmids are not self-transmissible; because they lack the *tra* genes. Nevertheless these small plasmids can infect other cells through a variety of means. In a process called **mobilization,** a

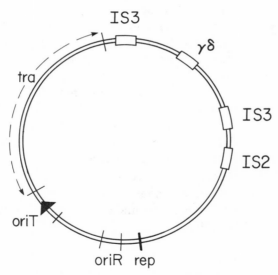

Fig. 8.2 Map of the F plasmid. The major operational regions are indicated: DNA replication (*ori*, the initiation site of replication; *rep*, genes encoding DNA replication), transfer (*tra*, genes encoding F pili synthesis, conjugation, and surface exclusion), and insertion sequences (IS, $\gamma\delta$).

small plasmid can cross over the conjugational bridge elaborated under the direction of a large transmissible plasmid. The ability to be mobilized is determined by specific genes and sites (mob) in the small plasmid. In addition, some nontransmissible plasmids seem to be readily packaged into bacteriophage particles and thereby can be transduced into other cells (see Chapters 9, 10, and 11).

Insertion Sequences and Transposons

A fundamental feature of plasmids that distinguishes them from bacterial chromosomes is the variety and frequency of genetic rearrangements that occur within these small genetic elements. Some of the rearrangements can be explained by $recA^+$-dependent recombinations discussed in the previous chapter. Others involve novel mechanisms mediated by insertion sequences (IS) and transposons (Tn).

F PLASMID INSERTION MECHANISMS

As mentioned in Chapter 7, in addition to the genes for conjugal transfer and replication, the F plasmid has a region with four insertion sequences (one of IS2, two of IS3, and one of $\gamma\delta$) that provides recombination points at which the plasmid can integrate into the E. coli chromosome to yield an Hfr strain. An IS2-mediated integration is shown in Figure 8.3. There exist numerous copies of IS2 in the chromosome of E. coli, so when one of these copies lines up with the IS2 of the F plasmid, a general recombination reaction catalyzed by the $recA^+$-coded protein can occur. As a result, the F plasmid is inserted into the chromosome. Typically the Hfr cells thus arising in an F^+ population constitute about 1 per 1,000 cells.

Hfr cells can arise from an F^+ population with no $recA^+$-coded protein, although only at a frequency of about 1 to 10^6. These are formed through the mediation of the $\gamma\delta$ sequence in the F plasmid which can presumably pick any region of the bacterial chromosome and cause F to integrate (Fig. 8.4). During this process of integration, the $\gamma\delta$ sequence duplicates itself such that it flanks the F. If subsequently these two $\gamma\delta$ sequences recombine with one another, in a reversal of a reaction similar to the one shown in Figure 8.3, the F would be excised from the chromosome, leaving behind one $\gamma\delta$. Thus

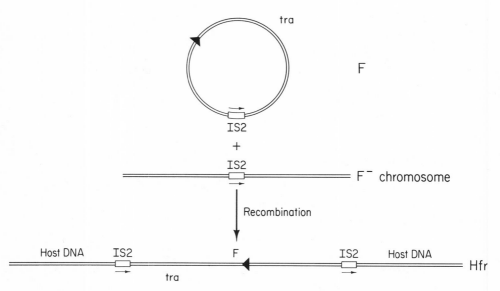

Fig. 8.3 F integration by IS2 homology. An IS2 in F (top) pairs with an IS2 in the chromosome by virtue of their homology. (An intermediate of this recombination reaction involves complementary pairing between the appropriate strands of opposite duplexes.) A crossover reaction is catalyzed by the recA protein. The result is the incorporation of a copy of F into the host DNA, making it an Hfr strain. Replication of any remaining F plasmids becomes repressed.

the net result of this insertion and excision is to transpose a copy of $\gamma\delta$ from F into the *E. coli* chromosome.

GENETIC REARRANGEMENTS IN R PLASMIDS

The number of drug-resistance determinants found in R plasmids is variable and generally not stable, especially for the large plasmids. Thus, continued culturing of a bacterial line carrying the plasmid R6 (Fig. 8.5) in the absence of antibiotics would give segregants that carry plasmids with *tet* but not *cml, kan-neo, str,* and *sul.* The most likely cause of the loss is a deletion generated by a recombinational event between the two IS1 sequences (Fig. 8.6). Conversely, one can imagine the formation of new multiresistance plasmids occurring when the small circle inserts into another plasmid. Such a process is analogous to the formation of an Hfr from an F$^+$, as shown in Figure 8.3.

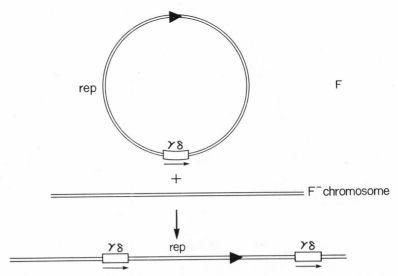

Fig. 8.4 F integration by $\gamma\delta$-mediated fusion. The $\gamma\delta$ insertion sequence in F (top) has the ability to integrate the plasmid into any site on the *E. coli* chromosome at a low frequency. The recA protein and extensive homology between the two genomes are not necessary. By a mechanism that remains mysterious, $\gamma\delta$ is duplicated during the integration process so that the F segment becomes flanked by two such sequences in the integrated state.

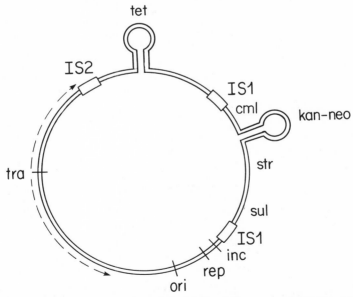

Fig. 8.5 Map of the large R plasmid R6. Three operational regions are apparent: transfer (*tra*), DNA replication (*ori*, *rep*), antibiotic-resistance (*cml*, *kan-neo*, *str*, *sul*), and insertion-sequence region (IS2, IS1, IS1).

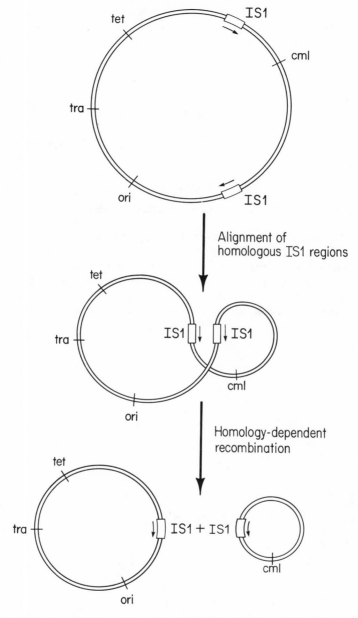

Fig. 8.6 Deletion in R6 mediated by IS1. After the two IS1 sequences align according to their homology, a crossover reaction catalyzed by the recA protein will cause the separation of two circles. Because *ori* is contained in only one of the two circles, subsequent cell growth will lead to the loss of the smaller circle.

Many drug-resistance genes can be rapidly amplified on the plasmid by means of **gene duplication.** For example, when cells containing R6 plasmids are challenged with excessive doses of chloramphenicol, the surviving cells will have plasmids with two or more copies of the cml gene, which consequently increase drug resistance. Such gene duplication is mediated by $recA^+$-dependent recombination that exploits the homology between the insertion sequences. Recall that each growing cell usually has more than one copy of the plasmid. Thus when the small circular DNA with the cml is formed, as in Figure 8.7, it may subsequently recombine with another wild-type R6 plasmid across the homologous IS1 region to give a duplication where the two cml genes are separated by an IS1. The net result is the transposition of the cml gene from one plasmid to the other. This $recA^+$-dependent duplication occurs at a relatively high frequency; perhaps a few percent of the plasmids at any given time carry such duplications even in the absence of special selective pressure.

A transposition reaction of the cml gene can also be mediated by the IS1 sequences in a way that is independent of $recA^+$. This is because insertion sequences can spontaneously transpose on their own. (They were initially discovered as the causative agents of insertion mutations; see Chapter 6.) However, such a transposition reaction is rare, perhaps found in one per 10^7 cells. But more importantly, a preexisting IS1 at the recipient site is not required. The occurrence can be demonstrated in a population of cells cured of R6 (for example, by treating them with acridine orange which selectively interferes with plasmid replication), and then exposed to chloramphenicol. The rare resistant clones that arise can be shown to contain the cml gene in the bacterial chromosome flanked by the IS1 sequences that are oriented with the same polarity (Fig. 8.7). Two IS1 sequences need not be in direct orientations (as they are in R6) but may be inverted with respect to one another and still mediate transposition. This reflects a general property of insertion sequences: two insertion sequences that flank any gene will enable that gene itself, along with the two insertion sequences, to transpose. This is why such a composite of genetic elements is called a **transposon.**

Nearly all drug-resistance genes are part of transposons. Figure 8.5 shows that the tet and kan-neo genes are each flanked by identical sequences known to have IS characteristics. The tet transposon is called Tn10, whereas the kan transposon is called Tn903; in these

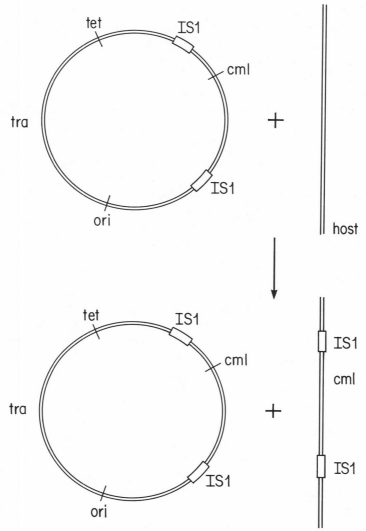

Fig. 8.7 Transposition of *cml* mediated by IS1.

two examples the flanking insertion sequences are inverted in orientation (hence shown as a stem structure).

GENETIC ORGANIZATION AND CONTROL OF TRANSPOSONS

Traits other than antibiotic resistance, such as toxigenicity, may be conferred by transposons. All transposons have a number of features

Fig. 8.8 Genetic organization of Tn5, a transposon conferring resistance to kanamycin. This resistance gene is flanked by two insertion sequences, in inverted orientation. The insertion sequence on the right (IS50R) codes for two proteins; one (coded by *tnp*) is needed for Tn5 to undergo transposition and the other (coded by *inh*) inhibits transposition. The two proteins are encoded within the same sequences of DNA by overlapping genes (see Chapter 10).

in common, however. They appear to carry genes that are needed for transposition (Fig. 8.8), as well as genes that repress or modulate the frequency of transposition. This is a necessary feature, since a transposon that continually implants itself into new sequence sites would eventually cause a lethal mutation resulting from a random insertion. Thus a transposon carries two signals: one promoting and the other inhibiting movement. The frequency and the conditions of insertion are determined by a balance between these signals.

The term "transposition" may be somewhat misleading in describing the process catalyzed by a transposon. The appearance of the sequence at a new location is not necessarily the result of moving the sequence out of its original position; rather the original sequence may serve as a template for copying itself. The copy is "transposed" while the original is conserved. This was shown for transposition of both $\gamma\delta$ and *cml*, as described above. Mechanistically unrelated, but necessary to balance the propensity of these elements to duplicate themselves, is the not yet understood ability of the bacterial cell to delete transposons.

Antibiotic Resistance

Plasmid-carried resistance is probably much more common in natural environments than resistance produced by spontaneous mutation of the bacterial chromosome, because of differences in the mechanisms conferring resistance. An R plasmid commonly brings resistance to a drug by carrying a gene encoding an enzyme that specifically catalyzes the inactivation of the drug. For example, resistance to streptomycin or neomycin is achieved by phosphorylating or adenylylating these drugs. In contrast, the chromosomal mutations that provide drug resistance usually alter an essential drug target site in the bacterium. For example, spontaneous streptomycin resistance usually arises by alteration of the gene for a ribosomal subunit preventing binding of the drug; neomycin resistance usually arises by alteration of the gene for an inner membrane transport protein involved in neomycin transport. Changes in functions of the target proteins are often slightly deleterious.

Not all plasmid-determined resistance to antibiotics is caused by inactivation. Tetracycline resistance is conferred by a gene coding for an inner membrane protein that actively transports tetracycline *out* of the cell. There is a case of an R-determined sulfonamide-insensitive enzyme in the biosynthetic pathway for folate, and a case of R-determined erythromycin resistance in staphylococci involving methylation of adenine residues of the 50S ribosomal RNA.

The most widespread resistance of therapeutic concern is probably that against the penicillins. In this instance, plasmid-determined resistance involves a gene that elaborates an enzyme (called β-lactamase) that cleaves the β-lactam ring of the penicillins. In contrast, spontaneous resistance in mutants isolated from a pure culture of *E. coli* usually involves one of the structural genes for peptidoglycan synthesis (from the pathway shown in Fig. 2.7).

EPIDEMIOLOGY OF PLASMIDS

The origin of the resistance genes in R plasmids has been subject to much speculation. In the case of chloramphenicol, the underlying mechanism for the resistance is the production of a specific acetylase. Acquisition of immunity by a mutation which causes the appearance of this kind of inactivating enzyme has never been observed. Yet chloramphenicol acetylase must have evolved through natural selection. It is known that R plasmids existed before the use of antibiotics in medicine, for they have been found in bacterial cultures preserved by freeze drying in vacuum before the widespread use of antibiotics. This is not surprising since antibiotics are natural products released by many kinds of microorganisms. It seems likely that detoxifying enzymes coded by R plasmids coevolved as a protective mechanism.

The ready transfer of the genes for resistance throughout the bacterial world is easily accounted for by the properties associated with R plasmids. Conjugal transfer explains how the R plasmids can spread from cell to cell. Broad host ranges ensure their distribution among different species of bacteria. For example, one particularly "promiscuous" plasmid, RP4, has been found in salmonellas, vibrios, shigellas, pseudomonads, agrobacteria, rhizobia, and other species of bacteria. The formation of plasmids with new combinations of genes is easily accounted for by the recombinations that plasmids can undertake and the association of many plasmid-borne traits with transposons.

Even though the R plasmids occur naturally, their current widespread distribution has clearly resulted from antibiotic use by humans. For example, hospital burn units that must use large amounts of various antibiotics to combat skin infection are one of the richest sources of R-plasmid-containing bacteria. But medical applications are not the only use for antibiotics; about half of all antibiotics produced in the United States are given to animals, not only to treat, but also to prevent disease and to promote growth of the animals. There is a rising suspicion that R plasmids multiplying in the bacteria of these animal populations pose a serious health problem, because intestinal flora can be interchanged between host species.

Questions

8.1. In pure culture, *E. coli* will give rise to kanamycin, tetracycline, and low levels of ampicillin resistances at frequencies of 10^{-10}, 5×10^{-11}, and 5×10^{-7} per generation, respectively. With what frequency would one expect to find mutants simultaneously resistant to all three antibiotics?

8.2. In R6, the two IS1 sequences are in direct orientation (see Figs. 8.5 and 8.6). What would be the result of a recombination event between the two IS1 sequences if they were in an inverted orientation?

8.3. A plasmid has the gene order shown in Figure Q8.3. (a) How many gene orders are possible after a single recombination event between the IS sequences? (b) How many gene orders after two?

8.4. When *E. coli* with the plasmid R6 is grown in the presence of high concentrations of chloramphenicol, cells with plasmids carrying dupli-

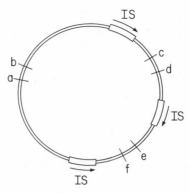

Fig. Q8.3

cations of the *cml* gene will be selected. When the genetic structures of the mutant and parent plasmids are compared, it is found that the duplication occurred through an IS1 as shown in Figure Q8.4. Such duplication is thought to arise by recombination and replication. Draw at least three pathways showing all the structures of the plasmids in each pathway.

8.5. Strain H of E. *coli* enterotoxigenic for humans, is assayed for its virulence in infant rabbits. Intraluminal inoculation of 10^5 viable cells results in pronounced distension of the small intestine after 18 hours and more than 10^{11} of strain H cells are recovered from the gut after rinsing by a standardized procedure. Repeated laboratory cultivation of strain H produced strain H-P, which can no longer cause diarrhea. In a similar assay, only about 10^7 H-P cells are recovered from the rabbit gut. Assays of enterotoxin shows strains H and H-P to produce the same amount. Antiserum raised against strain H and adsorbed with excess strain H-P still agglutinate strain H. The treated antiserum blocks colonization by strain H without killing the bacteria. In contrast to strain H, strain H-P does not produce pilus-like structures that react with untreated antiserum and lacks a plasmid containing 10^5 base pairs. The data indicate that intestinal colonization of the human enterotoxigenic E. *coli* depends on specific pili which are controlled by a plasmid. (a) How can the observation be explained in terms of the fate and properties of the plasmid? (b) Does the plasmid in question code for the pili? How might one prove it? Does it code for anything else?

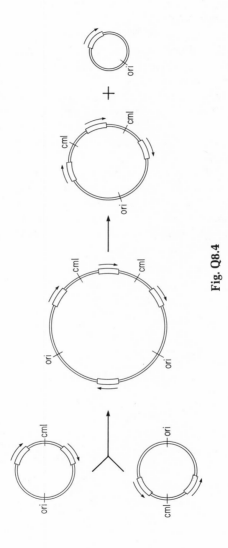

Fig. Q8.4

Further Reading

Bennet, P. M., and M. H. Richmond. 1978. Plasmids and their possible influence on bacterial evolution. In *The Bacteria: A Treatise on Structure and Function*, vol. 6, ed. L. N. Ornston and J. R. Sokatch. New York: Academic, pp. 2–62.

Broda, P. *Plasmids.* 1979. San Francisco: Freeman.

Calos, M. P., and J. H. Miller, 1980. Transposable elements. *Cell* 20:579–595.

Clark, A. J., and G. J. Warren. 1979. Conjugal transmission of plasmids. *Annual Review of Genetics* 13:99–125.

Falkow, S. 1975. *Infectious Multiple Drug Resistance.* London: Pion.

Foster, T. J. 1983. Plasmid-determined resistance to antimicrobial drugs and toxic metal ions in bacteria. *Microbiological Reviews* 47:361–409.

Kleckner, N. 1981. Transposable elements in prokaryotes. *Annual Review of Genetics* 15:341–404.

Konisky, J. 1982. Colicins and other bacteriocins with established modes of action. *Annual Review of Microbiology* 36:125–144.

O'Brien, T. F., J. D. Hopkins, et al. 1982. Molecular epidemiology of antibiotic resistance in *Salmonella* from animals and human beings in the United States. *New England Journal of Medicine* 307:1–6.

9

General Properties of Bacterial Viruses

VIRUSES are subcellular parasites that can multiply only inside living host cells. They use the energy sources, biosynthetic machinery, and even much of the replicative functions of the host cell, providing only the genetic information for their own replication. These genes code for proteins that replicate the viral genome, proteins that have structural or catalytic roles in assembling new infectious particles, and regulatory proteins that direct the overall process of viral reproduction.

The first description of viral action in bacterial cultures was made by F. W. Twort in 1915. Felix d'Herell gave the name **bacteriophage** to this bacteriolytic agent which he isolated from dysentery bacillus. (Today the term is conventionally shortened to **phage.**) He demonstrated that **lysis** (destruction) of cells in a phage-infected culture produced more phage, and that the lytic phage substance passed through filters which retain bacteria (Fig. 9.1). It is now known that in nature there exists a ubiquitous pool of phages and that probably every species of bacteria can act as a host for one or more viral species. However, not all viral infections are lethal to the bacterial cell; some so-called **temperate phages** can enter into symbiotic association with host bacteria, frequently introducing new properties to the infected cell.

Phages have been the object of intensive study over the past forty years, resulting in a wealth of information. Much of what is known about the nature of mutations, gene expression, and macromolecular assembly was originally learned from bacterial viruses. This knowl-

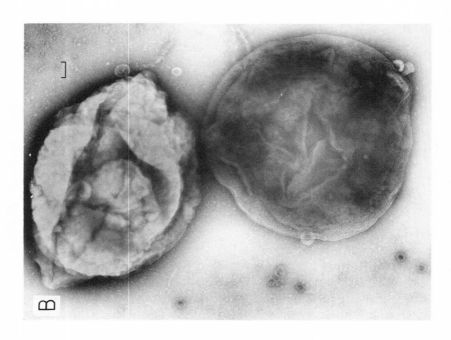

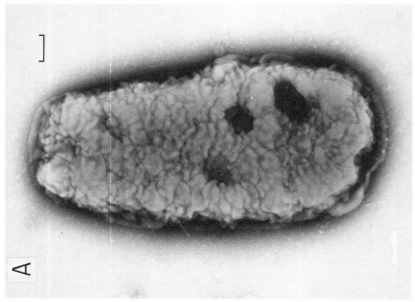

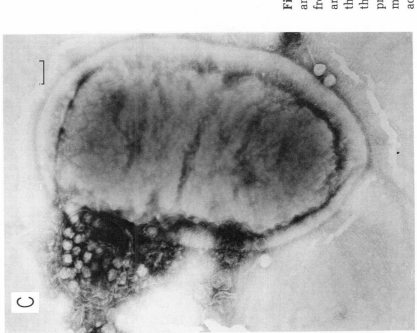

Fig. 9.1 Transmission electron micrographs of normal and lysed *E. coli* cells. (A) Typical uninfected cell taken from a culture in mid-exponential growth. (B) Cell lysed artificially with chloroform. (C) Infected cell caught in the midst of early lysis stage. Note the extrusion through the broken cell membrane of a vesicle containing many progeny viruses about to be released into the environment. All samples were negatively stained with uranyl acetate. Scale: 0.1 μ. (By R. N. Goldstein.)

edge in turn provided a foundation for the study of animal virology. Recently, phages have attracted new interest because of their importance to recombinant DNA technology, where they are used as carriers (or **vectors**) of cloned genes from both eukaryotic and prokaryotic organisms (see Chapter 12). Specially engineered variants of the phage called λ have been used in E. coli for the cloning of eukaryotic genes to produce such proteins as growth hormones, insulin, and interferons.

Of particular significance to clinical bacteriology is the finding that many temperate phages code for genes specifying toxins, thus endowing the host bacteria with particular advantages. Diseases such as diphtheria, scarlet fever, and botulism are caused by bacteria that have been infected with such phages (Fig. 9.2). The narrow host

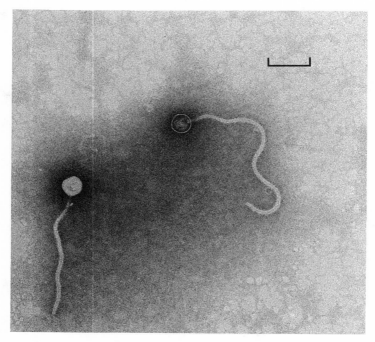

Fig. 9.2 Transmission electron micrograph of Corynebacterium phage responsible for conversion of an otherwise nonpathogenic bacterium into a producer of diphtheria toxin. The sample was negatively stained with phosphotungstic acid. Scale: 0.1 μ. (By R. N. Goldstein.)

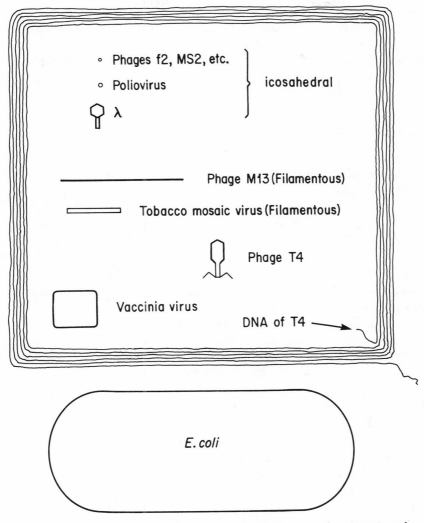

Fig. 9.3 Schematic diagram demonstrating relative size of various icosahedral or filamentous phages in comparison with animal viruses, a plant virus, and an *E. coli* cell. The compactness of DNA packing in a virus is brought into perspective by the relative length of the T4 genome shown as a framework.

specificity (or host range) of some phages is exploited in diagnostic tests to identify pathogens. This technique (called **phage typing**) is routinely used in hospital laboratories for differentiating the many strains of typhoid bacilli (some of which are responsible for typhoid fever) and staphylococci, for example.

Structure and Composition

Phages in their inert extracellular form are compact particles consisting of the genome surrounded by protein subunits that form a shell or **capsid** (also referred to as "head" or "coat"). Figure 9.3 indicates the range in size and morphology of viruses that attack bacteria, animals, and plants. The capsid acts not only as a protective container for the genome, but some of its protein subunits also play a role in packaging the genome, adsorbing (binding) to the host cell, and injecting the genome into the bacterial host. There is considerable similarity in the basic types of capsid structure among viruses of eukaryotic and prokaryotic cells.

A phage capsid encloses a single copy of the viral genome, consisting usually of one molecule of either double-stranded DNA (as in $E.$ $coli$ phages λ, T4, T7, P2, and P4), single-stranded DNA ($E. coli$ phages ϕX174 and M13), double-stranded RNA ($Pseudomonas$ phage 6), or single-stranded RNA ($E. coli$ phages Qβ, R17, and f2). (The genome of $Pseudomonas$ phage 6 consists of three different double-stranded RNAs. This allows recombination by reassortment.) The chromosome of a virus may account for up to 50% of its total mass. A phage chromosome may be extremely small (such as the genome of $E. coli$ phage R17, which is approximately 3,600 bases in length and contains four genes) or relatively large (such as the genome of $E. coli$ phage PB51, which is approximately 2.5×10^5 bases in length and contains over 240 genes).

There are three basic forms of phage capsid (Fig. 9.4). One type has

Fig. 9.4 Electron micrographs of phages illustrating the three basic capsid forms. (A) Phage P2, which has a "complex" capsid consisting of an icosahedral shell attached to a contractile tail with tail fibers. (Phosphotungstic acid stain; by R. N. Goldstein.) (B) Phage ΦX174, which has a "simple" icosahedral capsid only. (Phosphotungstic acid stain; by R. N. Goldstein.) (C) Phage fd, which has a filamentous capsid. (Courtesy of R. C. Williams.) Scale: 0.05 μ.

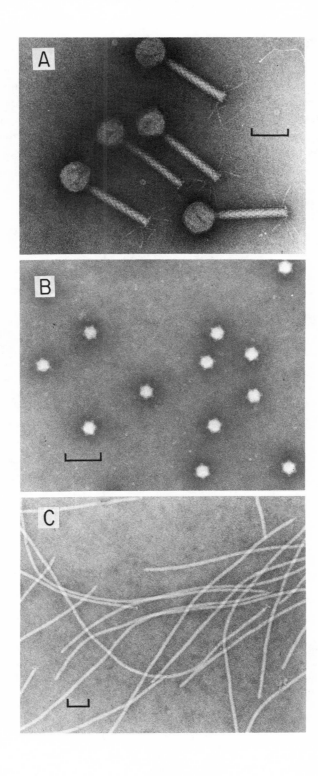

protein subunits arranged in icosahedral symmetry (like a geodesic dome) and are called spherical viruses. The simplest known contains only 60 protein subunits, whereas more complex ones contain multiples of 60. A second type of capsid can be described as an icosahedral structure to which are attached auxillary protein components: internal and external collars (plates), a cylindrical tail tube (sometimes

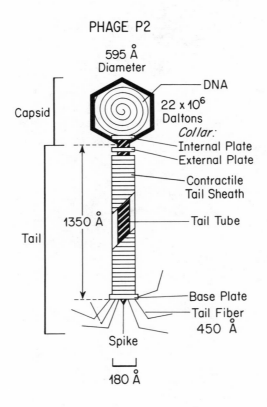

Fig. 9.5 Schematic diagram of phage P2, a typical complex double-stranded DNA phage that infects *E. coli*. The icosahedral capsid contains some 540 copies of the 36,000-dalton major capsid subunit protein. The contractile tail sheath is composed of about 200 copies of a 46,000-dalton subunit protein, and the tail tube consists of some 320 copies of a 19,000-dalton subunit protein. (Modified from D. Shore, G. Deho, J. Tsipis, and R. N. Goldstein. 1978. Determination of capsid size by satellite bacteriophage P4. *Proceedings of the National Academy of Sciences, USA* 75:400–404.)

surrounded by a contractile sheath), a base plate, tail fibers, and spike (Fig. 9.5). *E. coli* phages T4, P2, and λ and *Corynebacteria* phage β are of this type. A third type of capsid is filamentous in structure. It is essentially a hollow cylinder with helically arrayed protein subunits. Some filamentous phages specifically attack male bacterial cells by using the sex pili as receptors. Some filamentous phages employ the host cell's plasma membrane as cover.

Adsorption and Transfer of the Genome

The interaction between phages and bacteria is highly specific, requiring an attachment protein on the phage and a receptor protein on the host cell. On phages, the protein subunit essential for adsorption varies. In the case of *E. coli* phage T4, the initial engagement is by the six tail fibers with tryptophan as an essential cofactor. For *Salmonella* phage P22, a base plate with extended spikes is used for binding. Filamentous phages bind by one of their tubular ends, and some simple icosahedral phages, such as ϕX174, are believed to use a minor capsid protein located at one of the twelve vertices.

On the host cell's surface, many types of accessible molecules may serve as specific phage receptors. Proteins exploited by phages as receptors generally have important roles for the normal functioning of the cell. For example, in *E. coli*, phage λ binds to the outer membrane protein for maltodextrin transport; ϕ80 and T5 bind to the outer membrane protein for ferrichrome transport; T2 binds to the outer membrane porin protein (forming channels for the diffusion of hydrophilic molecules); MS2 and f2 (both filamentous RNA phages) and T7 (an icosahedral DNA phage) bind to the F pilus for cellular conjugation.

More than one specific interaction may be necessary for irreversible engagement between virus and host. In the case of T4, it appears that following the binding of the individual tail fibers to the receptors on the outer cell membrane, the tail itself anchors to a different receptor (Fig. 9.6). In the case of the filamentous phages that adsorb to the tip of the pili, it seems that the fastened pilus retracts, bringing the complete phage down to and then through the outer membrane.

The transfer of phage genetic material into the cytoplasm of the host cell likewise calls for a complex interplay between the receptor

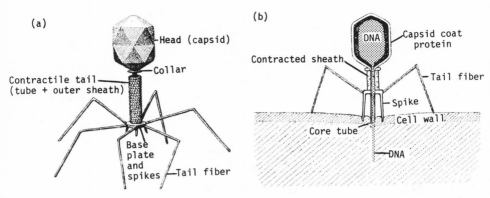

Fig. 9.6 Adsorption of phage T4. (a) The basic structures. (b) The relationship between structure and function. Upon adsorption, the sheath contracts, and the tail tube is driven into the cell wall. Contraction of sheath protein (induced by tail fiber and core interaction with receptors on cell surface) then triggers release of DNA from the capsid through the hollow core tube into the cell. (Adapted from W. B. Wood and R. S. Edgar. 1968. *The Molecular Basis of Life.* San Francisco: Freeman, p. 155.)

and phage structure, a process that requires energy input. This complexity can be readily appreciated in the gram-negative cell, whose several envelope layers must be penetrated by the phage genome.

Contact of phage T4 with the outer membrane receptors initiates conformational changes in the phage structure which cause the tail sheath to contract, forcing the hollow inner tube into the cell. These events also change the conformation of proteins in the base plate, tail tube, collar, and possibly the head, allowing release of the genome through the tube into the cell. The injection of the DNA is driven by proton motive force maintained by the bacterial cell. Phage λ, another virus with a tail, does not contract its tail sheath during DNA injection. In the above examples, the empty capsid remains attached to the cell surface after the DNA enters the cell. In contrast, filamentous DNA phages of *E. coli*, such as fd, seem to enter the cell by being drawn into the inner membrane of the cell envelope while being uncoated; the DNA is released intracellularly as the coat protein dissociates into subunits which remain in the membrane.

Expression of Viral Genes

Although a wide variety of strategies for viral reproduction have been discovered (see Chapter 10), the sequence of events is conventionally described as a four-stage process.

(1) Following the entry of the viral chromosome, the genes expressed early code for proteins which are involved in phage genome replication and which modify the cellular machinery so that the synthetic capacity of the cell is subverted to the reproduction of the phage. These early gene products are rarely found in the completed phage.

(2) The remaining genes that specify "late" functions are then expressed to initiate the synthesis of the various protein subunits of the capsid, which are assembled into intermediate structures.

(3) The genomes are packaged within capsid and ancillary proteins, and mature viruses appear.

(4) The phage progeny is released, usually as a result of host-cell disintegration which is hastened by a phage-coded lysis enzyme, such as a lysozyme (Fig. 9.7).

The intracellular multiplication of a virulent phage is sometimes described as **vegetative growth.**

Infection by **temperate** phages is also generally lytic, but occasionally a state of dormancy results. Many kinds of phages achieve this dormancy by integrating their DNA into the host chromosome, a process referred to as **lysogeny** (see Chapter 11). Infection by certain filamentous phages can also **persist** without being lethal to the bacteria. Instead, the host cells intermittently extrude progeny phage (Fig. 9.8).

Growth of Lytic Phage in Solid Bacterial Culture

Lysis of bacteria by phages can be conveniently visualized in solid cultures. In this procedure Petri plates containing a layer of nutrient agar are seeded with bacterial cells at an invisibly low density. After several hours of growth, the bacterial colonies become confluent.

Lytic Infection

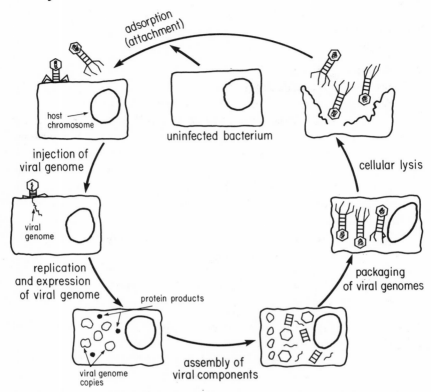

Fig. 9.7 The lytic response. The phage binds to a specific receptor on the outer membrane and injects the viral genome. Expression of viral genes redirects host metabolism, resulting in the replication of viral DNA and production of capsid subunits. Capsids assemble and then package the new copies of viral genome. Cell lysis eventually occurs, releasing infectious progeny into the environment.

This confluent bacterial population is often referred to as a **lawn.** When a single lytic phage infects a bacterium at the time of seeding, a **plaque** (or clear zone) will eventually appear in the lawn. This happens because lysis of the originally infected cell liberates many progeny phages, which in turn infect and subsequently lyse surrounding cells. A plaque, which can reach several millimeters in

Persistent Infection

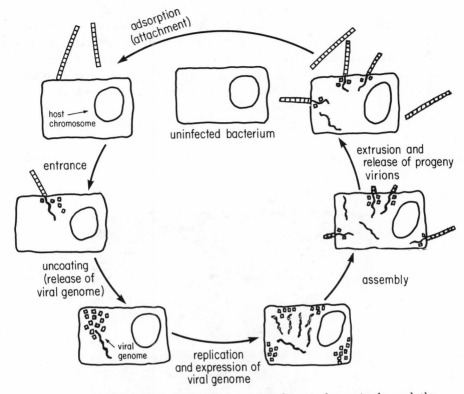

Fig. 9.8 Persistent infection. The filamentous phage is drawn in through the cell envelope and uncoating of the viral genome occurs. Replication of viral DNA and synthesis of new capsid subunits proceed at an attenuated pace. Mature viruses are extruded through cell membranes into the environment without causing cell lysis.

diameter, may contain from 10^4 to 10^8 viruses, depending on the kind of phage and the suitability of the host cells. Plaques generally do not increase indefinitely in size, because most phages only replicate in growing cells.

Plaque morphology and size are characteristic for particular phage–host systems. Thus plaques may be described as large, minute, turbid, ragged-edged, and so on (Fig. 9.9). Some of these characteristics are used in phage genetics to identify mutants.

Plaque formation is a standard method for assaying the infective titer of a phage suspension (adapted also to titer animal viruses), as illustrated in Figure 9.10. The number of phages can also be determined by physical or chemical means, such as their visualization by electron microscopy or measurement of nucleic acid content. The

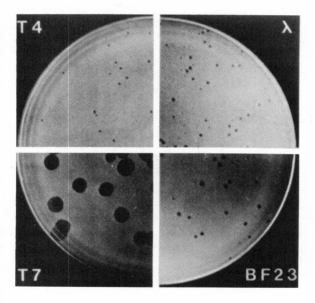

Fig. 9.9 Characteristics of plaques produced by four different phages. Note the difference in plaque sizes, T7 producing the largest and T4 the tiniest. The size of a plaque may depend on the burst size and the speed of viral multiplication. Alteration in plaque size for a given species of phage may indicate a mutation that affects some aspect of the lytic cycle. (From R. E. Glass. 1982. *Gene Function: E. coli and Its Heritable Elements.* London: Croom Helm, p. 207.)

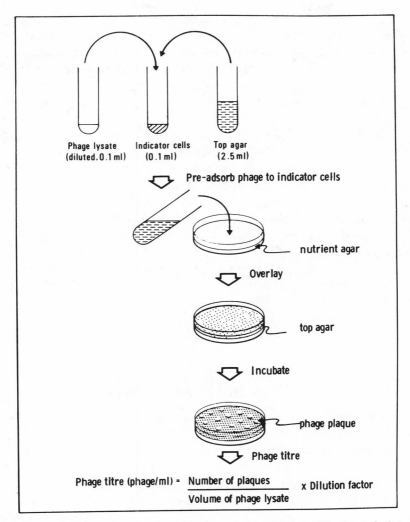

Phage lysate (diluted, 0.1 ml) Indicator cells (0.1 ml) Top agar (2.5 ml)

Pre-adsorb phage to indicator cells

nutrient agar

Overlay

top agar

Incubate

phage plaque

Phage titre

$$\text{Phage titre (phage/ml)} = \frac{\text{Number of plaques}}{\text{Volume of phage lysate}} \times \text{Dilution factor}$$

Fig. 9.10 The plaque assay procedure. An appropriate dilution of phage suspension is mixed with a suspension of indicator (permissive) cells in great excess. After 10 minutes for adsorption to take place, the mixture is evenly diluted with a molten solution of low-concentration agar. The entire content is poured over the surface of hardened nutrient agar. The top agar layer solidifies after cooling. After incubation for a period of several hours, the number of plaques are counted. (From R. E. Glass. 1982. *Gene Function: E. coli and Its Heritable Elements.* London: Croom Helm, p. 208.)

biological assay (that is, plaque formation) seldom gives as high a value as a physical or chemical assay. It seems that some bacterial cells for physiological reasons interact with a phage in an abortive way and thus will not support its growth. (This is true even when the cellular defense against foreign DNA by specific nucleases is not a significant factor; see Chapter 12.) Also, some phages are defectively assembled, even under the best of circumstances. It therefore follows that the **efficiency of plating,** defined as the ratio of plaque-forming units (pfu's) to phage particles, is usually less than one. For most phages the ratio is quite low, perhaps 10% or even less.

Finally, plaques are also used as the primary technique for isolating viruses from natural sources. Our knowledge that almost all bacteria have phages that attack them comes from the simple experiment of mixing a pure culture of the bacterium with a possible source of phage (sterilized by filtration or chloroform treatment to eliminate living cells) and plating the mixture for plaques. The potential source of phage would be any place where the bacteria grow. Thus, for *E. coli* one might use raw sewage.

Growth of Lytic Phage in Liquid Bacterial Culture

In experiments with phages the ratio of pfu's to the number of host cells in a suspension is set at the time of infection. This ratio is called the **multiplicity of infection** (MOI). A high MOI (for example, 5:1) assures that almost all cells have one or more infectious particles attached to them. Multiple infection is possible because there are usually many receptor sites per cell. In contrast, with a low MOI (for example, 1:100) one can be reasonably sure that each individual cell is infected by no more than one phage.

If susceptible bacteria growing exponentially in a liquid culture are exposed to phages at a high MOI, one cycle of phage multiplication will bring about lysis of the whole culture. This process is described as **one-step growth.** The time required may be 15–60 minutes, depending on the phage, the host bacteria, the growth medium, and the temperature. Basic features of phage infection were first revealed by this kind of growth experiment. The protocol is as follows:

(1) Mix an exponentially growing culture of about 10^8 cells/ml with phage to give an MOI of 5.
(2) Allow 10 minutes for adsorption, and then add sufficient antibodies to inactivate unadsorbed phages. (Further activity of the adsorbed phage is not affected since attachment to the host cell membrane has already occurred.) This will give a culture of infected cells.
(3) Dilute the culture 1,000-fold into more medium, incubate at 37°C, and take samples at intervals for phage titering. Note that in such a situation the total number of **infective centers** will be the sum of infected bacteria and free phages, if any. An infected cell will produce only a single plaque, because the progeny of phages ultimately liberated will be confined to a single focus by the surrounding agar.

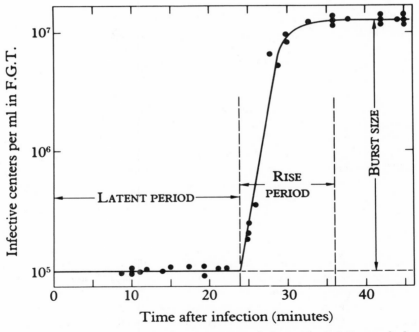

Fig. 9.11 The one-step growth experiment of phage T4. F.G.T. stands for first growth tube and represents the 1,000-fold dilution of the original infected culture. (From G. Stent and R. Calendar. 1978. *Molecular Genetics: An Introductory Narrative.* 2nd ed. San Francisco: Freeman, p. 304. After A. H. Doermann. 1952. *Journal of General Physiology* 35:645, by copyright permission of The Rockefeller University Press.)

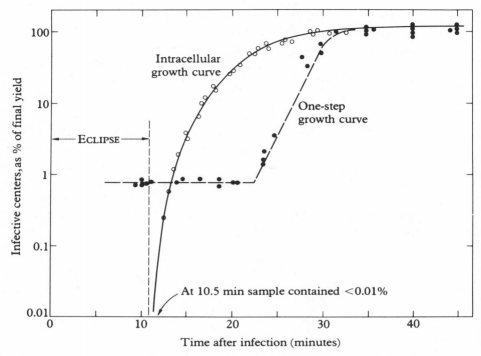

Fig. 9.12 The artificial lysis experiment demonstrating the kinetics of intracellular phage growth. (From G. Stent and R. Calendar. 1978. *Molecular Genetics: An Introductory Narrative.* 2nd ed. San Francisco: Freeman, p. 306. After A. H. Doermann. 1952. *Journal of General Physiology* 35:645, by copyright permission of The Rockefeller University Press.)

Fig. 9.13 Electron micrographs of the intracellular events accompanying the reproduction of phage T4. (0 minute) The *E. coli* cell has the normal morphology, with the typical nucleoid (less-dark parts) of cells at exponential growth. (5 minutes) Enzymatic hydrolysis of the cellular DNA has occurred, as shown by the nuclear disruption. Nucleotides of the host DNA are being used for synthesizing phage DNA under the control of phage genes of early functions. (15 minutes) The first infectious phages appear. It takes about 7 minutes to assemble and mature a particle. Inset shows prehead structures with little or no DNA content. (30 minutes) New phage particles appear at a rate of about 5 per minute. (This is a thin cross section; therefore not all of the particles are shown.) (Courtesy of Dr. Michel Wurtz.)

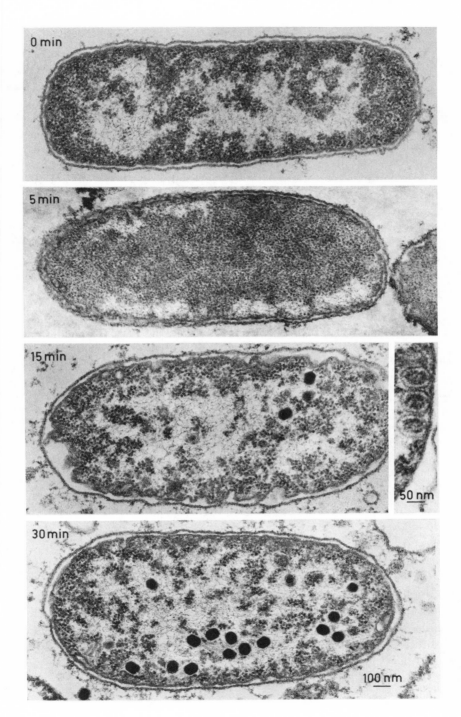

0 min

5 min

15 min

50 nm

30 min

100 nm

The results of such an experiment are depicted in Figure 9.11. For the first 24 minutes, the pfu's remain constant and are equal to the number of initially infected cells. The infection is said to have entered the **latent period.** The fact that the pfu's all represent infected cells can be shown by centrifuging a portion of the culture before titration and then assaying the supernatant fluid (where only free phages remain). From 24 to 36 minutes the pfu's in the culture begin to rise rapidly **(rise period)** until a final plateau is reached at 36 minutes, after which there is no further increase. The latent period represents the time required for the first cell in the culture to lyse, thereby releasing new phage progeny into the medium. The rise period covers the span during which more and more of the infected bacteria lyse, and the final plateau is reached when all infected cells in the culture have lysed. The **burst size** ($10^7/10^5 = 100$) corresponds to the average number of phage progeny produced per infected bacterium. This value can vary from 2 to 1,000, depending on the particular phage–host system and physiological conditions. Lysis of the bacterial culture causes clearing of the suspension.

A variant of the one-step growth experiment can reveal more about phage development. Here the same protocol is carried out except that samples taken at different times are treated with chloroform to lyse the cells chemically before titering mature phage particles by plaque formation. Figure 9.12 shows that the infectivity associated with the original parental phage is completely lost at the outset of the infection. This **eclipse period** lasts about 10 minutes. Next, a constantly increasing number of infective phages begin to make their intracellular appearance. This increase continues until the final titer reaches the same value as in a parallel experiment in which no artificial cell lysis is used. During the eclipse period, the injected phage genome is being replicated and phage capsid subunits are synthesized, but these elements are not yet infective. Intracellular packaging of the first newly replicated phage genome by the capsid gives rise to the first infective progeny and marks the end of the eclipse. Displayed in Figure 9.13 are electron micrographs of thin sections cut through T4-infected *E. coli* cells at various times after initial infection. The eclipse period can be seen to end at about 14 minutes. Figure 9.14 shows schematically a lytic cycle of T4 development.

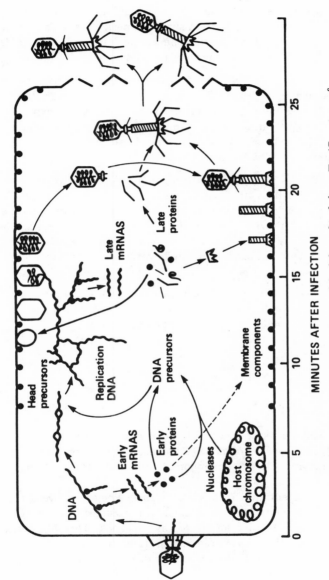

MINUTES AFTER INFECTION

Fig. 9.14 Schematic representation of the life cycle of phage T4. (Courtesy of Professor C. K. Matthews.)

Questions

9.1. Although wild-type phage T4 can carry out the entire infectious process in E. coli at both 30° and 42°C, phage mutants that no longer do so at 42°C can be isolated. Design simple experiments to show whether temperature-sensitive mutations affect (a) adsorption, (b) DNA injection, (c) DNA replication, (d) phage assembly, or (e) release.

9.2. 10^{10} T4 phages are added to 10^{10} sensitive E. coli in suspension and allowed to adsorb. When this culture is titered for viable colony-forming units a value of 3×10^9 per ml is obtained. If the experiment is repeated using phage λ, the number of viable cells is found to be 7×10^9. (a) In these two cases how did the surviving cells avoid lethal infection? (b) Why the difference?

9.3. Phages T4 and T7 are virulent. (a) If one infects a growing bacterial culture (10^9 cells) with T4 at 10 phages per cell, the culture will clear within an hour because of cell lysis. However, after overnight incubation, the culture will become turbid. Why? (b) If a culture is infected simultaneously with T4 and T7, clearing occurs as before, but now when the lysed culture is incubated overnight, no turbidity will arise. Why?

9.4. Two derivatives of phage X, each with a mutation in a different structural gene, failed to yield progeny in a mixed infection. Suggest at least four models to explain the failure of complementation.

Further Reading

Adams, M. 1959. *Bacteriophages*. New York: Interscience.

Barksdale, L., and S. B. Arden. 1974. Persisting bacteriophage infections, lysogeny and phage conversions. *Annual Review of Microbiology* 28:265–299.

Cairns, J., G. S. Stent, and J. D. Watson. 1966. *Phage and the Origins of Molecular Biology*. New York: Cold Spring Harbor Laboratory.

Denhardt, D. T., D. Dressler, and D. S. Ray, eds. 1978. *The Single-Stranded DNA Phages*. New York: Cold Spring Harbor Laboratory.

Stent, G. 1963. *Molecular Biology of Bacterial Viruses*. San Francisco: Freeman.

———, ed. 1960. *Papers on Bacterial Viruses*. Boston: Little, Brown.

Stent, G., and R. Calendar. 1978. *Molecular Genetics: An Introductory Narrative*. San Francisco: Freeman.

Tikhonenko, A. S. 1970. *Ultrastructure of Bacterial Viruses.* New York: Plenum.

Warren, R. A. J. 1980. Modified bases in bacteriophage DNAs. *Annual Review of Microbiology* 34:137–358.

10

Lytic Development of Phages

THE CHAIN of events beginning with transfer of the phage genome into the cell and ending with the death of the host is called **lytic development**. As might be expected, such a process is tightly regulated, though its mechanisms vary considerably from one species of phage to another.

Replication of Phage Genomes

As was mentioned in Chapter 9, encapsidated phage genomes come in several types (Fig. 10.1). The smallest are the single-stranded linear chromosomes of RNA phages coding for only a few proteins. The chromosome of ϕX174 is a small circular single-stranded DNA molecule. The other examples shown are double-stranded linear DNA chromosomes. T4 chromosomes are not all alike; the genes from a given end of the chromosome appear in cyclic permuted order. Furthermore, all of these chromosomes are terminally redundant (that is, the base sequences at the two ends are identical). The chromosome of T7 also has terminal redundancy, but the base sequence from one end to the other is invariant. The λ chromosome has short single-stranded linear ends twelve bases long which are complementary to each other such that they can anneal to give a circular structure. Variations in the basic chromosomal structure of these phages with double-stranded DNA genomes reflect different strategies of their replication and packaging.

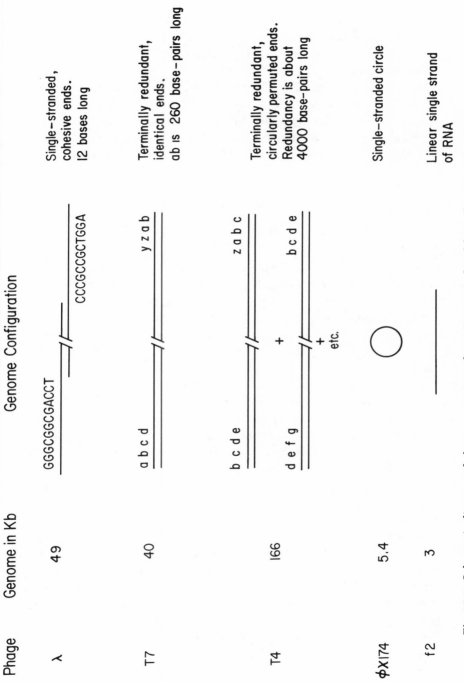

Phage	Genome in Kb	Genome Configuration	
λ	49	GGGCGGGCGACCT CCCGCCGCTGGA	Single−stranded, cohesive ends. 12 bases long
T7	40	a b c d y z a b	Terminally redundant, identical ends. ab is 260 base−pairs long
T4	166	b c d e z a b c + d e f g b c d e + etc.	Terminally redundant, circularly permuted ends. Redundancy is about 4000 base−pairs long
φX174	5.4	◯	Single−stranded circle
f2	3	——	Linear single strand of RNA

Fig. 10.1 Schematic diagram of phage genome configurations as isolated from the respective viruses.

Though the genome of many phages is packaged as a linear molecule in the infectious particle, most of the chromosomes are replicated as circular molecules. Replication of the double-stranded chromosome of DNA phages proceeds in a manner similar to that of the plasmids as outlined in Chapter 8. DNA synthesis is initiated at the ori of the chromosome, and the intermediate structures have θ forms in electron micrographs. But unlike the plasmids, replication of a single infecting genome of a lytic phage exponentially increases the copy number from one to many hundreds, ending with cell lysis; there are no diffusible replication inhibitors like those made by plasmids.

PHAGE λ

Upon entry into the host cell the cohesive ends join by hydrogen bonding to form a circle which is covalently sealed by DNA ligase. The genome undergoes a few rounds of duplication as circles (θ replication) and then switches to σ replication; that is, it changes from two forks moving in opposite directions to one fork moving unidirectionally. Despite the radical difference in the appearance of the intermediate forms emanating from these two processes (Fig. 10.2), fork elongation reactions may be the same in the two cases. The θ form can be converted to a σ form simply by cutting a circular parental strand at one fork, joining the 3' cut end in register with the 5' end of the daughter strand copied from the complementary parental circle, and permitting the other fork to continue elongation. When this is done the active remaining fork can traverse a number of times around the uncut parental strand (acting like a rolling circle), generating one genome length per cycle (Fig. 10.2). This is like the unidirectional replication seen with the F plasmid during conjugal transfer (Fig. 7.4), with the exception that the growing tail in phage replication is double-stranded. The long double strands containing multimeric units of chromosome equivalents are called **concatemers.** These tandem genomes are cut by a terminase in a staggered way to generate the cohesive ends; the terminase also plays a role in genome packaging. Phage λ encodes two proteins that help activate its ori and presumably direct the host replication enzymes to catalyze the synthesis of phage DNA.

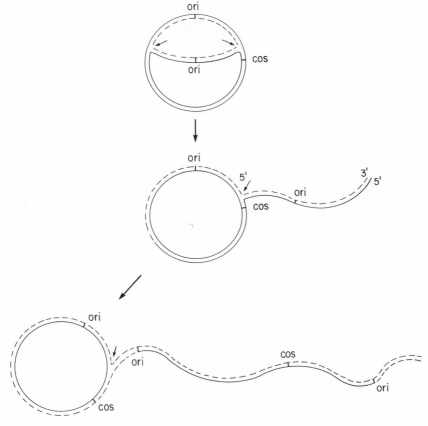

Fig. 10.2 Unidirectional rolling-circle DNA replication producing a σ intermediate form. The result is a string of concatenated daughter molecules which are cut at *cos*, producing staggered, cohesive ends.

PHAGE T4

One of the largest and most complex of viruses is phage T4 with about 200 genes. In rough outline, the replication cycle of the T4 genome proceeds much as that of λ. The linear DNA molecule injected into the cell circularizes probably by a recombination reaction between the two identical sequences at the end of the chromosome, resulting in the loss of the redundant bases. These circles undergo θ replication early in infection and switch to a rolling circle (σ) replication later in infection to produce long concatemers. Unlike

λ, T4 codes for essentially all of the 20 proteins that are involved in its DNA replication cycle. Many of these play roles similar to the counterpart cellular enzymes. This apparent duplication of effort is probably necessitated in part by the unusual composition of T4 DNA with 5-hydroxymethylcytosine replacing cytosine. Furthermore, the modified cytosine is glucosylated. It should not be surprising, therefore, that many phage-specified enzymes are involved in nucleotide metabolism. The presence of unusual bases in the viral genome is not rare. The odd structure of the DNA might be used by the phage for offensive or defensive roles (see below and Chapter 12).

PHAGE T7

The chromosome of T7 never circularizes in the infected cell. Replication of the genome begins at a point 17% from one end and proceeds in both directions, producing linear intermediates with "eye" and "Y" forms (Fig. 10.3). The progeny molecules form conca-

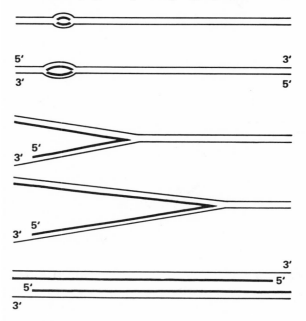

Fig. 10.3 Bidirectional replication of linear T7 DNA giving rise to intermediate "eye" and "Y" forms. (Adapted from A. Kornberg. 1974. *DNA Synthesis*. San Francisco: Freeman, p. 270.)

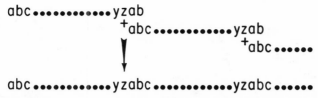

abc •••••••••••yzab
⁺abc •••••••••••yzab
⁺abc ••••••

abc •••••••••••yzabc •••••••••••yzabc ••••••

Fig. 10.4 Concatemer formation by recombination of T7 genomes. The terminal redundancy is used for homologous crossover.

temers, probably by recombination at the identical DNA sequences at the ends of the unit chromosomes (Fig. 10.4). This phage has about twenty genes, at least eight of which are involved directly or indirectly with DNA replication: a protein kinase that inactivates host RNA polymerase by phosphorylation, a new RNA polymerase, a DNA polymerase with three accessory proteins (binding protein, helicase, and primase), a DNA endonuclease, and a DNA exonuclease. The two nucleases contribute to the break-up of host DNA, providing nucleotides for new DNA synthesis; in fact, 99% of the nucleotides incorporated into phage progeny are derived from the host chromosome.

PHAGE ϕX174

Single-stranded DNA phages pose a unique problem of replication: the entering genome does not carry a template for its own synthesis. This DNA molecule, conventionally designated (+), is copied by host enzymes to give a complementary strand, designated (−). With the completion of the (−) strand, a closed circular duplex is first produced which is called a **replicative form.** The DNA at this point is physically indistinguishable from a small plasmid. The replicative form is then copied in much the same way as plasmids are to give more duplex circles.

Late in infection, a cut in the (+) strand by a phage protein gives a new intermediate. The (−) strand then serves as template for synthesis of new (+) strands. This causes displacement of a (+) strand, reminiscent of rolling-circle replication. The expelled (+) strands are not replicated into (−) strands but rather are packaged to become progeny phages (Fig. 10.5).

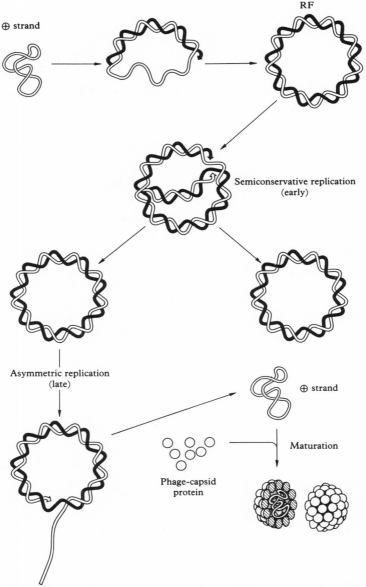

⊕ strand

RF

Semiconservative replication
(early)

Asymmetric replication
(late)

⊕ strand

Phage-capsid
protein

Maturation

Fig. 10.5 A general scheme for intracellular development of phage ΦX174. Completion of the (−) strand with the incoming (+) strand as template is followed by θ replication which finally switches to σ replication. (From G. Stent and R. Calendar. 1978. *Molecular Genetics: An Introductory Narrative.* 2nd ed. San Francisco: Freeman, p. 326. Copyright © 1978 by W. H. Freeman and Company. All rights reserved.)

DNA Packaging

Late in infection DNA is packaged into capsid structures. For the large double-stranded DNA phages with replicated DNA in the form of long concatemers, there are two general processes for packaging the genome. One is described as **site-to-site.** In this process a phage-encoded endonuclease recognizes a specific base sequence in the DNA and cleaves the concatemer there. For example, if a site yzab occurs once per genome length in the concatemer (ab---yzab---yzab---yz), and a cut is made between z and a, each capsid will contain a single genome segment ab--yz.

A consequence of site-to-site packaging is that the amount of DNA to be contained in the phage capsid is determined more by the specific cutting sites located in the DNA sequence than by the overall size of the molecule. Thus, phage λ, which packages its DNA by the site-to-site process, may produce altered virions carrying from 30% less to 12% more DNA. (This is important for specialized transduction and the cloning of DNA, discussed in more detail below.) Staggered cuts at specific sites of the concatemer account for the single-stranded complementary ends of the λ chromosome in the phage particle. In contrast, packaging of the T7 DNA by staggered cuts followed by synthetic extension of the 3' short ends has been proposed as a mechanism to account for the identical sequences with direct repeats (terminal redundancy): ab--yzab.

The other process of packaging unit genomes from concatemers is described as **headful.** In this case the capsid is filled to capacity with DNA and the cutting is not determined by any specific site in the concatemer. Phages T4 and P1 are examples. Since the capsid can generally hold more than a single genome length of DNA, the chromosomes are terminally redundant. The headful packaging of the T4 chromosome containing 103% of the full genetic information explains not only this terminal redundancy but also the circularly permuted gene order (abcd . . . yzab, cdef . . . zyabcd, and so on).

Generalized Transduction

Phage capsids that package their DNA from concatemers by the headful mechanism have the potential of enclosing any DNA that may be present in the bacterial cell late in infection, including host

DNA if any remains. In the case of T4 infection, degradation by the T4 nucleases of the host DNA is already complete, because it is unprotected by glucosylation. In contrast, *E. coli* phage P1 and *S. typhimurium* phage P22, which do not elaborate such nucleases, do package host DNA; lysates of these phages will contain normal viruses as well as viral particles that carry headful fragments of host chromosomes.

These pseudo-viruses can adsorb and inject their DNA into sensitive cells; the result is not an infection but the introduction of a fragment of bacterial genome. This is known as **generalized transduction** and provides a way for the exchange of genetic information between different bacterial strains (Fig. 10.6).

Just as in the transformation of the pneumococcus by pieces of DNA (see Chapter 7), fragments of bacterial chromosome introduced by general transduction cannot replicate. However, they can be incorporated into the host chromosome by recombination. For example, if a lysate of phage P1 previously grown on wild-type *E. coli* is used to infect a $recA^+,lacZ^-$ mutant strain, transductants that can grow on lactose will arise from the treated bacterial population. No transductants will arise if the recipient is $recA^-,lacZ^-$. Transductants that acquired the $lacZ^+$ marker are also likely to have incorporated several contiguous genes from the donor. Generalized transduction is thus an important technique for genetic mapping and bacterial strain construction.

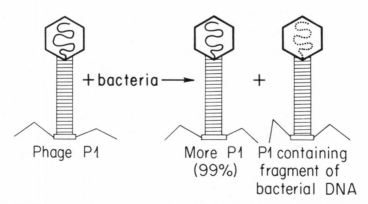

Phage P1   + bacteria ⟶   More P1 (99%)   +   P1 containing fragment of bacterial DNA

Fig. 10.6 Generalized transduction by phage P1. Headful DNA packaging makes the phage apparatus a vehicle for transferring surviving bacterial DNA to other host cells.

Exogenous genes will not be incorporated into the newly infected cell's chromosome if there is a lack of homology between the foreign DNA and the native genome. The imported genes may still function, but in a limited way. Suppose in this situation a his^+ fragment has entered a his^- recipient. The merodiploid is able to make histidine and grow without an external supply, but the his^+ DNA fragment, not being able to replicate, will be inherited by only one daughter cell. Provided the fragment is not soon degraded, the cell containing the his^+ DNA will continue to divide to form a microcolony by linear rather than exponential growth. The cells in a microcolony are called **abortive transductants;** they are mostly daughters of the single cell that inherited the his^+ fragment. Without the his^+ DNA, growth will cease when the mRNA and its encoded enzymes are degraded or diluted out.

Regulation of Gene Expression

In many respects, the expression of phage genes resembles that of bacterial genes: transcription may be under positive or negative regulation, and post-transcriptional and post-translational control may be operative. However, there are aspects of phage gene expression that have no counterparts in bacteria or are much more difficult to study in cells. One is the timing of events in the growth cycle. In many cases, expression of genes controlling late functions depends on the occurrence of early events. Because most phage genomes are so much smaller than bacterial chromosomes, it has been feasible to study the temporal order of gene expression from the moment the phage genes are introduced into the cell.

In general, the control of phage gene expression is at the level of transcription, as is true for bacterial genes. Since RNA polymerase plays a central role in bacterial gene expression, it is not surprising that a number of phage control mechanisms tamper with the activity of this enzyme. There are cases of its modification or replacement altogether by a phage-coded protein. The result is to alter the program of gene expression so phage DNA instead of host DNA is transcribed or one set of phage genes rather than another is transcribed. Early versus late functions are often regulated in this way.

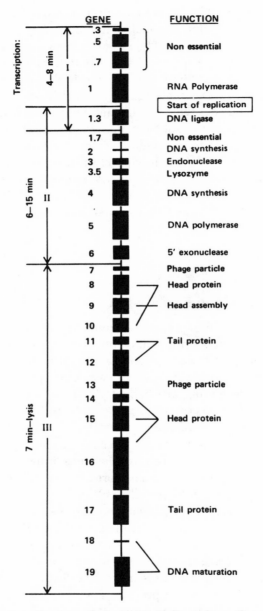

Fig. 10.7 The genetic map of phage T7. Indicated are the gene numbers, their estimated sizes and functions. The regions transcribed early, the subsequent stages of the phage cycle, and their duration are also shown. (From A. Kornberg. 1974. *DNA Synthesis*. San Francisco: Freeman, p. 268.)

PHAGE T7

All the genes of phage T7 are coded on the same strand, and their topographic order corresponds to their temporal order of expression: early, middle, and late (Fig. 10.7, groups I, II, and III, respectively). The genes of group I are transcribed by the host RNA polymerase. (Evidently the promoters of the group II and III genes do not bind to the host enzyme.) One of these gene products is a simple RNA polymerase, and another codes for a protein kinase that inactivates host RNA polymerase. Transcription of the phage genes in groups II and III requires the new T7 RNA polymerase and proceeds to the right-hand end. Thus, the order of gene expression in this phage is set by the fact that T7 early genes are transcribed by the host enzyme, whereas the mid- and late gene transcriptions are dependent on phage-coded enzymes and factors.

PHAGE T5

This double-stranded DNA phage of 1.2×10^5 base pairs has an unusual mechanism for preventing the simultaneous expression of all of its genes. The phage genome contains several specific breaks in one strand of the DNA; the initial injection proceeds only up to the second break (at 8%). Transfer of the rest of the genome into the cell happens four minutes later, after two new DNA-binding proteins are synthesized from the initially injected piece.

PHAGE N4

Unlike the other bacterial DNA viruses described, the virulent E. coli phage N4 has the unusual characteristic of not depending on the host RNA polymerase at all for phage expression; concentrations of rifampin that totally inhibit host RNA synthesis do not prevent phage-dependent RNA synthesis. A rifampicin-resistant RNA polymerase is carried by the phage and enters the cells with the DNA. This situation is analogous to the virus-associated polymerases in some animal viruses.

SATELLITE PHAGE P4

This "helper-dependent" double-stranded DNA phage of 1.1×10^4 base pairs codes for about twelve gene products. In contrast to other phages, it does not code for its own capsid proteins. In the absence of a helper phage which provides these gene products, the P4 genome either integrates into the host chromosome or maintains itself in plasmid fashion. However, in the presence of an integrated or co-infecting helper, such as temperate phage P2, P4 undergoes development as a lytic phage. P4-coded regulatory gene products derepress, activate, and modulate the expression of the helper genome, thus causing the helper to supply the necessary proteins for the assembly of P4 particles and for cell lysis.

Of particular interest is that regulatory proteins encoded by P4 do not simply substitute for those of the helper P2. Rather, the P4 regulatory proteins modify the phenotypic expression of P2 so that it produces a new type of small capsid capable only of packaging the smaller P4 genome. One of the P4 proteins responsible for this process appears to interfere with the normal transcription termination of P2 genes encoding capsid proteins.

PHAGE ϕX174

An astonishing feature of this small phage with a single-stranded DNA genome (5.4×10^3 nucleotides) is that six of its genes overlap

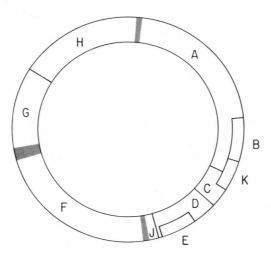

Fig. 10.8 The genetic map of phage ΦX174. There are a total of ten genes, with extensive overlapping of the genes in the lower right quadrant of the map. Note gene K shares coding regions with two other genes, A and C; and gene A shares coding regions with two other genes, B and K. The total genome has 5,386 bases which have been sequenced in F. Sanger's laboratory. Shaded regions are not coding for proteins.

but are frame-shifted relative to each other. Thus overlapping regions can code for proteins with no common amino acid sequence (Fig. 10.8). This ingenious arrangement allows great economy in the use of DNA for coding.

RNA Phages

Phages with RNA genomes are among the smallest viruses known; their chromosomes are generally about 3,000 bases long. In spite of the small size, their cycle of growth can be divided into several stages in a manner similar to those of the larger and more complex viruses. All *E. coli* RNA phages (R17, MS2, $Q\beta$, and so on) are of a single class that can infect only host cells that carry the F plasmid, since the F pilus serves as the surface receptor. A single-stranded linear RNA chromosome is enclosed in an icosahedral shell composed of 180 coat-protein subunits and one A (or maturation) protein subunit.

The single-stranded RNA isolated from phage particles serves as mRNA. As with single-stranded DNA, by convention the strand with base sequences corresponding to mRNAs is designated as (+) and its complementary copy as (−). The genome encodes at least four proteins: the coat protein, the A protein, a replicase (RNA polymerase) subunit, and a lysis enzyme. Because the number of codons needed to store the information for these four proteins exceeds the length of the chromosome, portions of the genome must code for more than one polypeptide: coding economy is achieved by having the lysis gene overlap on one end the region coding for the C-terminal portion of the coat protein and on the other end the ribosome-binding site and the N-terminal portion of the replicase gene (Fig. 10.9). The codons for the lysis gene are in a different reading frame from the codons in the coat protein and replicase genes.

In spite of the simplicity of genome organization in RNA phages, the synthesis of proteins is temporally regulated much like that of the more complex phages already described. The coat protein is first synthesized after infection and is followed almost immediately by the appearance of the replicase. These may be considered the early proteins. After a delay, the synthesis of A protein commences while the synthesis of the replicase ceases. Coat-protein synthesis continues late into the infection, and the cell ruptures as a result of the synthesis of the lysis protein.

Control of protein synthesis is exerted at the level of translation.

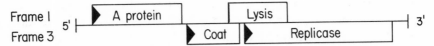

Fig. 10.9 Organization of the four genes in the single-stranded chromosome of RNA phage MS2 serving also as mRNA. Boxes demarcate coding regions of the genes. Genes indicated above the line use reading frame 1 (which begins at the first base), and genes indicated below the line use reading frame 3 (which begins at the third base). An arrow designates the position of the ribosome-binding site for each gene and the direction of translation (ribosomes can translate only 5′ to 3′ along mRNA). Since RNA synthesis proceeds also in the 5′ to 3′ direction, during replication of the new (+) strand the first ribosome-binding site to be synthesized is that for the A protein.

Essential to the mechanism of switching from early to late protein synthesis is the fact that the phage RNA inside the cell is not actually in the linear form as represented by the genetic map in Figure 10.9 but rather assumes a compact folded structure much like the polypeptide chain in a globular protein. Folding is brought about by regions in the mRNA molecule that are complementary to each other, causing the formation of intramolecular duplexes (snapbacks). The biological importance of this folding is the occlusion of all of the ribosome-binding sites except the one for the synthesis of the coat protein. Once translation of the coat protein begins, however, the threading of its mRNA through the ribosomes physically disrupts the folding and exposes the ribosome-binding sites for the translation of the replicase, thus activating its synthesis.

Synthesis of the maturation protein takes place after RNA replication commences. RNA-dependent RNA replication is not a usual process in prokaryotic or eukaryotic cells. The polypeptide coded by the phage replicase gene has to be associated with three different host cell proteins in order to acquire the replicase activity. The host subunits of this enzyme complex are the elongation factors, Ts and Tu, and a ribosomal protein, all of which are components of the normal host translational machinery. The RNA replicase complex copies the (+) strand brought in by the infecting phage, giving a full length (−) strand. In turn, the (−) strands serve as templates for production of new (+) strands, and so forth (Fig. 10.10). By the time of lysis, perhaps 5,000 to 10,000 (+) strands will have been synthesized. There will be an excess of (+) strands over (−) strands, because late during infection (+) strands are packaged into coat structures to

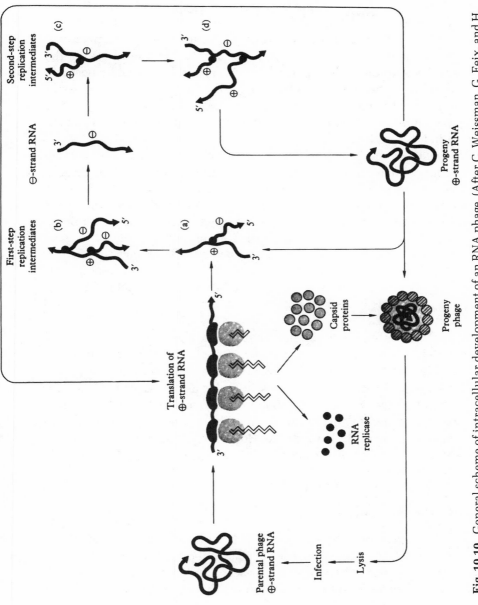

Fig. 10.10 General scheme of intracellular development of an RNA phage. (After C. Weissman, G. Feix, and H. Slor. 1968. Cold Spring Harbor Symposia on Quantitative Biology 33:83.)

make mature phage particles, leaving behind free (−) strands to continue serving as templates for the synthesis of more (+) strands.

It is during the copying of the (+) strand from the (−) strand that the synthesis of A protein is observed. Presumably this is because the ribosome-binding site of the A gene is exposed for translation only in the nascent (+) strand. In fact, each nascent (+) strand causes the synthesis of only one molecule of A protein before the ribosome-binding site becomes blocked again as the replication process passes the critical region. The topography of the genome thus insures a 1:1 stoichiometric synthesis of the A protein and the (+) RNA strand, which is the ratio seen in virions.

Later in the infection, the coat protein accumulates to a level sufficient to bind significant stretches of the (+) strand; these covered sequences include the ribosome-binding site of the replicase gene. When this happens, further replicase synthesis is terminated. Thus the coat protein also acts as a translational repressor of the replicase gene. The mechanism of activating translation of the lysis gene remains unknown. Possibly a conformational change in the (+) strand induced by the binding of coat proteins initiates the synthesis. Curiously, the lysis gene apparently has no ribosome-binding site; the ribosomes loaded on an upstream gene might continue to complete the task.

In brief, through the use of overlapping genes, multifunctional proteins, internal base pairing of the single-stranded chromosome, and the host proteins in an adventitious and opportunistic way, the information coded in four genes suffice to assure an orderly procession of phage reproduction. Single-stranded viruses may be much more vulnerable to mutations and thus selectively pressured to maintain a small target genome. However, developmental organization seems not to be sacrificed on this account.

Phage Assembly

The complexity of the viral capsid reflects its multiple functions. The structure must: (1) compactly package and protect the genome from destructive agents such as nucleases, (2) attach and interact with specific receptor sites of the host bacterium, and (3) release the genome into the host cytoplasm. Simple self-assembly by continuous condensation of individual subunit proteins is inadequate for the development of such a complex structure. Instead, discrete component parts are preassembled before they combine to form the final structure. In many cases the process is aided by enzymes and proteins that serve as transient ramparts.

In the case of T4, all the structural proteins are synthesized simultaneously during late gene expression and three independent morphogenic pathways converge to constitute the mature virus (Fig. 10.11). The assembly therefore cannot be guided by the temporal order of subunit synthesis: instead, the program must be imposed by specific sequences of protein interactions. A number of proteins take part in this assembly without being incorporated themselves as structural components of the mature viral particle. These have been classified into three categories: scaffolding proteins, enzymes, and promoters of noncovalent subunit associations. The basic component structures can be revealed by electron microscopy of spontaneous lysates of cells infected with wild-type phage. These preparations contain not only completely assembled viruses but also distinct parts and larger precursors at different stages of assembly. From the kinds of intermediate forms found, it is apparent that the joining of up to 40 different components is not haphazard but follows a strict order. Furthermore, structural anomalies in some of the incomplete particles can be detected, which suggest that they are defective intermediates rejected from the morphogenic pathways. Thus the parallel assembly of the major structures (heads, tails, fibers, and so forth) probably provides a screening mechanism to ensure that the viral genome is eventually enclosed by a fully functional container. There is no evidence for feedback control of subunit synthesis by the finished components. The various proteins seem to be synthesized at fixed ratios, with tails and tail fibers in excess over heads.

The formation of the icosahedral T4 phage head well illustrates the intricacy of the assembly process. A major protein first condenses around a scaffolding matrix protein (later to be discarded) to form a core. Several different proteins are then introduced to elaborate a prohead shell. At least one host protein is involved in some way. The prohead assumes three successive forms which are brought about by the participation of at least three more different scaffolding proteins and a number of additional structural protein subunits. The mature prohead is then enlarged by proteolytic cleavages of inner and outer proteins by a phage enzyme before the genome is received.

If a single viral protein is absent or nonfunctional as a result of a

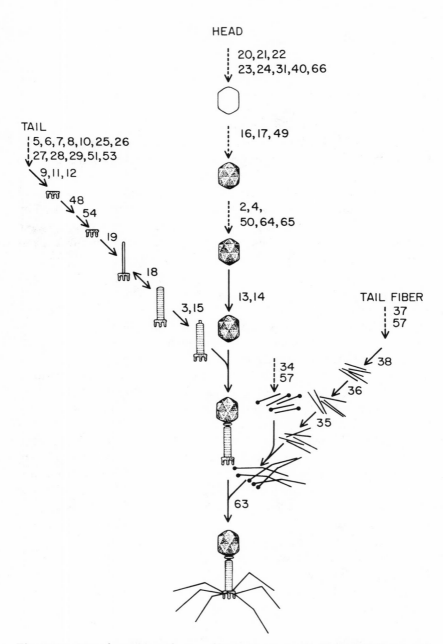

Fig. 10.11 Morphogenic pathway of bacteriophage T4. Tails, tail fibers, and heads (capsids) are each assembled through distinct pathways (involving one or more intermediate precursor structures). Numbers represent gene products of individual T4 genes involved in each step. (Adapted from W. B. Wood, R. S. Edgar, et al. 1968. *Federation Proceedings* 27:1160–1166. Courtesy of Professor W. B. Wood.)

mutation, a particular kind of intermediate structure will accumulate, and all proteins to be added in subsequent steps will remain in solution. Definitive characterization of the precursors accumulated by various kinds of mutants has been an important tool in unraveling the process of particle assembly. Phage T4 encodes at least 53 different gene products involved in its assembly.

Viral Genetics

Much of our knowledge about how phages function comes from studies of mutants. Since most phage gene products are necessary for the various stages of reproduction outlined above, the functions of these proteins can be most conveniently analyzed with the help of conditional mutants.

Temperature-sensitive (ts) and chain-terminating nonsense mutants (see Chapter 6) are commonly used. For example, one may screen for a phage that produces plaques at 30°C but not at 42°C. Such a phage will carry a mutation in an essential gene causing its protein product to be thermolabile. Or one can select phage mutants that grow in a bacterial strain with tRNA suppressor (sup) but not in strains without the suppressor $(sup°)$. Phages with nonsense mutations in essential genes are propagated in a sup host, whereas the lethal effect of a mutation can be studied by infecting these phages into a $sup°$ host.

Complementation tests similar in principle to those described in Chapter 7 can distinguish whether or not two mutations affect the same gene. Because two or more phages can infect the same cell, it is possible to introduce two different mutant chromosomes into the same cytoplasm (Fig. 10.12) under nonpermissive conditions. Inside the infected cell, if the two mutant genomes are defective in the same gene, then no progeny phage will be produced under the nonpermissive condition. On the other hand, if the defects are in different genes, then the function of the mutant gene in one genome may be supplied by the other genome, and vice versa; that is, complementation will occur. Ordinarily this will allow the complete replication cycle to proceed in that cell. The phages produced will still carry either of the original mutations and will not grow under nonpermissive conditions. Such experiments have shown, for example, that phage T7 requires at least 19 genes for its growth.

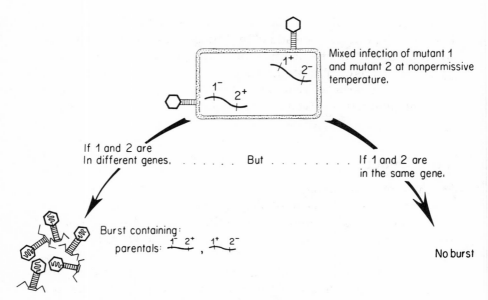

Fig. 10.12 A phage complementation experiment with temperature-sensitive mutants. Each phage has a mutation affecting a different essential gene (outcome given in the left pathway) or the same essential genes (outcome given in the right pathway).

Some phage genes, however, encode products that are not needed for growth in the experimental host, but have more subtle functions. Mutations in these functions, which may affect the size or shape of plaques, have been identified by their phenotype. The rapid-lysis (rII) mutants of phage T4 which form large plaques are one example. Some fundamental discoveries in molecular biology, such as the triplet nature of the genetic code, were made with these mutants, even though it was not known what protein was coded by the rII gene. The clear-plaque mutants of phage λ were critical in the studies of the lysogenic response (see Chapter 11). Another class of tolerable mutations affects host range. If one selects bacterial mutants resistant to a given phage, it is often possible to select phage mutants that are in turn able to form plaques on the bacterial mutant; mutations of this kind might entail the alteration of a phage protein needed for adsorption.

Questions

10.1. *E. coli* strain B supports the growth of the two closely related phages, T2 and T4. *E. coli* strain B/2 supports the growth of phage T4 but not T2 because this bacterium lacks the receptor for the adsorption of T2. When cells of *E. coli* strain B are co-infected with both T2 and T4, the progeny includes not only these two phages, but also products of **phenotypic mixing,** that is, phage particles with the genome of one phage and the capsid of the other. What results, in terms of plaque formation, would be expected from plating this mixed crop of phage particles on an initially invisible lawn of the following cells and incubating the agar dishes for a day: (a) *E. coli* strain B, (b) *E. coli* strain B/2, (c) a mixture of the two bacteria?

10.2. A mutant of *E. coli* resistant to bacteriophage BF23 is isolated. When the mutant is studied further, it is found to have the additional property of being unable to transport vitamin B-12 into the cell. Explain.

10.3. In a cell carrying a UAG (amber) suppressor mutation and infected with a T4 phage carrying an amber mutation, (a) compare the translation products of the mutant phage gene and the wild-type phage gene; and (b) explain why the phage burst size under these conditions can be up to 80% if the amber mutation is in an "early gene," but only 20% if the mutation is in a "late gene."

10.4. What is the biological significance of modification of host RNA polymerase as an early phage function?

10.5. Chain terminating nonsense mutations in the coat protein gene of an RNA phage are isolated. Those that map near the $5'$ end of the gene are defective not only for coat protein synthesis but also replicase synthesis. On the other hand, those that map near the $3'$ end of the gene not only synthesize replicase but produce more than the normal amount of it. How might the discrepancy be explained?

Further Reading

Blumenthal, T., and G. G. Carmichael. 1979. RNA replication: function and structure of Qβ-replicase. *Annual Review of Biochemistry* 48:525–548.

Earnshaw, W. C., and S. R. Casgens. 1980. DNA packaging by the double-stranded DNA bacteriophages. *Cell* 21:319–331.

Falco, S., R. Zivin, and L. B. Rothman-Denes. 1978. Novel template requirements of N4 virion RNA polymerase. *Proceedings of the National Academy of Sciences, USA* 75:3220–3224.

Geisselsoder, J., M. Chidambaram, and R. Goldstein. 1978. Transcriptional control of capsid size in the P2:P4 bacteriophage system. *Journal of Molecular Biology* 126:447–456.

Murialdo, H., and A. Becker. 1978. Head morphogenesis of complex double-stranded DNA bacteriophages. *Microbiological Reviews* 42:529–576.

Rabussay, D., and E. P. Geiduschek. 1977. Regulation of gene action in the development of lytic bacteriophages. *Comparative Virology* 8:1–96.

Richardson, C. C. 1983. Bacteriophage T7: minimal requirement for the replication of a duplex DNA molecule. *Cell* 33:315–317.

Sanger, F., A. R. Coulson, et al. 1978. The nucleotide sequence of bacteriophage ϕX174. *Journal of Molecular Biology* 125:225–246.

Wickner, W. 1983. M13 coat protein as a model of membrane assembly. *Trends in Biochemical Sciences* 8:90–94.

Wood, W. B. 1979. Bacteriophage T4 assembly and the morphogenesis of subcellular structure. *Harvey Lecture Series, 73*. New York: Academic.

Zinder, N., ed. 1975. *RNA phages*. New York: Cold Spring Harbor Laboratory.

11

Lysogeny in Temperate Phages

THE INFECTION of a sensitive cell by a lytic phage always leads to cell disruption and death. Infection by a temperate phage, on the other hand, has two possible consequences: the phage may undergo lytic development and kill the host bacterium; or, less frequently, the phage chromosome may become established nonvirulently in the cell. In this latter case, called the **lysogenic response,** the phage chromosome is said to enter a **prophage** state; the host cell survives but will replicate the prophage along with the bacterial chromosome so that all descendants inherit the prophage genome. Some kinds of prophages are propagated like independent plasmids, others as integral parts of the bacterial chromosome.

The prophage is genetically silent, because a complex repression system acts to prevent the expression of most of its genes. Nonetheless, a bacterial strain that carries the prophage can be induced to undergo lytic development, and this is one way by which such strains are recognized. A strain carrying repressed prophage is therefore called a **lysogen** (able to lyse). Aside from the latent ability to make phage, lysogens have another fundamental characteristic: they are no longer sensitive to infection by sibling phage. They are resistant in part because the repression system that prevents the prophage from expressing itself also prevents lytic growth of a new infecting phage of the same type.

Since infection of sensitive bacteria by a temperate phage causes most cells to lyse, these phages typically produce turbid plaques on a

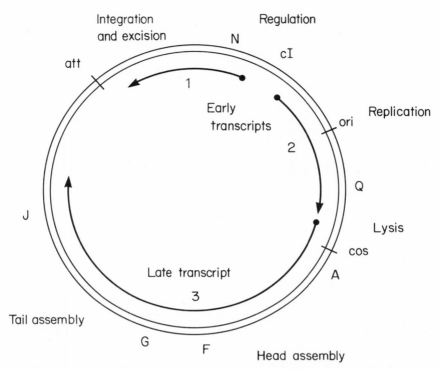

Fig. 11.1 Genetic structure of λ. The genetic map is organized with functionally related structural genes clustered together. Arrows inside the circle show major transcriptional units that act during lytic growth (see Fig. 11.2 for transcriptional units that regulate lysogeny). Genetic symbols are as follows: *cos* for the site where cohesive ends of the linear genome close to produce a circle; *att* for phage attachment site which interacts with the bacterial chromosome; *ori* for origin of DNA replication, N-cI-Q for region controlling gene expression, and A-J for the 20 genes determining the assembly of the phage.

lawn of the indicator bacteria. The turbidity results from the outgrowth of numerous microcolonies of lysogens within the plaque.

One might surmise that temperate phages are more complex than virulent phages, and indeed none of the temperate phages are very small. Phage λ is a temperate phage of *E. coli* strain K-12 and is probably the best understood complex virus, not just for lysogeny but for all aspects of its growth. Phage P22 is a temperate phage of the bacterium *Salmonella typhimurium*. There are a number of other phages whose interaction with the host cell differs in significant ways, but most of the following discussion will be concentrated on λ. The genetic organization is shown in Figure 11.1.

Gene Expression of Phage λ

When *E. coli* is infected with λ, the outcome is determined by at least six phage-encoded gene products that exert their influence on at least six DNA sites. To understand the regulatory mechanisms that lead to a lysogenic response in λ, it is necessary first to understand its lytic response, since that response must be completely repressed in order for lysogeny to occur. As in lytic development of virulent phages, transcription of the λ genes is regulated in a temporal order. Immediately after infection, the host RNA polymerase recognizes the promoters p_R and p_L and initiates transcription in the directions shown in Figure 11.2. After a few genes have been transcribed, the synthesis of mRNA is prematurely stopped at the ρ-dependent terminator sites t_R and t_L. These mRNAs are the early transcripts. The leftward mRNA then directs the synthesis of the N protein. N is a positive control protein that antagonizes transcriptional termination sites at t_R and t_L, allowing the RNA polymerase to read through these sites and transcribe the gene cIII on the left and the gene Q on the right. These products are the middle transcripts (1 and 2 shown in Fig. 11.1). Lytic growth requires only the rightward transcript. The Q protein is a positive control element required for transcription of the late region from p_Q (transcript 3); this region contains the information for the synthesis of the late proteins taking part in the viral assembly and cell lysis. In essence, lytic growth results from a cascade of activating events: free RNA polymerase causes the formation of N, N then acts to allow expression of Q, and finally the Q regulatory gene product promotes the expression of late morphogenic gene products needed for phage assembly and cell lysis.

As with some of the virulent phages described in Chapter 10, λ's early RNA synthesis is subsequently shut off during the course of lytic growth. The cro protein, encoded by the rightward early transcript, is responsible for this task and does so simply by binding to o_R and o_L, causing inhibition of further transcription from p_R and p_L. By then sufficient Q protein already has been synthesized, and transcription to form early mRNA is no longer necessary, since the Q

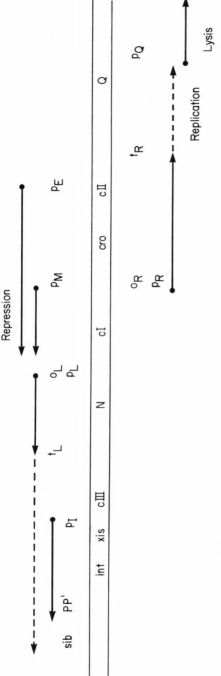

Fig. 11.2 Controlling region of λ. Genes for proteins are indicated between the two lines. Regulation sites indicated above the lines affect transcription leftward, whereas those indicated below the lines affect transcription rightward. This convention takes note of the fact that transcription in a given direction can proceed only from one DNA strand, since RNA polymerase can synthesize RNA only in the 5'-3' direction.

Genes

cI	encodes the λ repressor that binds to $o_L + o_R$, activates p_M
N	encodes the antiterminator that acts at t_L and t_R
cII	encodes activators of transcription from p_E (and p_I)
cIII	same activity as cII
cro	encodes the antagonist that acts at $o_R p_R p_M$ and $o_L p_L$ against the repressor
Q	encodes the activator of transcription of late genes
int	encodes integrase
xis	encodes excision enzyme

Sites

o_R, o_L	λ operators
p_R, p_L	early and middle promoters
t_R, t_L	ρ-dependent terminaters
p_Q	late promoter
p_I	int promoter
p_M	maintenance promoter
p_E	establishment promoter
sib	retroregulation of int gene
PP'	phage attachment site (designated att in Fig. 11.1)

regulatory gene product can now maintain expression of late genes involved in the lytic response.

MAINTENANCE AND ESTABLISHMENT OF THE LYSOGENIC STATE

How is the prophage state maintained once lysogeny has been established? In the prophage, the only gene transcribed is cI which codes for the λ repressor. The repressor binds to the operators o_R and o_L, blocking synthesis of the two early transcripts which are necessary for lytic development. On the other hand, the λ repressor acts as a positive control element for the transcription of its own gene from the promoter, p_M (promoter for maintenance). Thus when λ repressor binds to the o_R-p_R-p_M site, RNA polymerase will no longer transcribe p_R in the rightward direction; instead, it will transcribe p_M in the leftward direction, assuring an ample supply of the repressor. In essence, the cI protein acts to cause transcription at this complex promoter and regulatory site to reverse direction. Since the λ repressor is a diffusible product, it will attach to the o_R and o_L sites on any superinfecting λ chromosomes that enter a lysogen. This accounts for the specific immunity of lysogens. (Repression is phage specific.)

Because λ repressor is required for its own synthesis in the lysogen, the question of how repressor synthesis is first established must be raised. As in the lytic response, the rightward and leftward transcripts from p_R and p_L must be completely expressed. This leads to the synthesis of not only the N protein but also protein cII and subsequently protein cIII, which are positive control elements that activate the promoter, p_E (promoter for establishment). Transcription of this promoter is leftward and includes the repressor gene cI. Once the repressor is made, the maintenance promoter, p_M, is activated. We can summarize the lysogenic response as the culmination of a slightly different cascade of transcriptional events starting with transcription from p_R and p_L by free RNA polymerase, followed by N-prolonged transcription (synthesis of cIII protein is dependent upon N), continued by transcription of p_E activated by proteins cII and cIII, and ending with transcription at p_M activated by protein cI. At this point, the cI-coded repressor turns off both the establishment and the lytic chain reactions by binding to o_R and o_L and blocking p_R and p_L.

LYTIC GROWTH VERSUS LYSOGENY

Whether the response is lysogenic or lytic appears to be dependent on whether the cII-cIII-cI chain reaction or the N-Q chain reaction predominates. The cro protein is critically involved in this decision. As mentioned above, cro can bind to o_R to shut off rightward transcription from p_R. When this happens, transcription of cI from p_M is also shut off, that is, cro represses transcription of cI from p_M, thus preventing the repressor from maintaining its own synthesis. In short, the cI repressor and cro protein compete for binding at the o_R-p_R-p_M site. If the repressor prevails $early$ in infection, the lysogenic response occurs; if not, then the lytic response occurs. Exactly how this balance is tilted remains unknown, although it does appear that physiological conditions of the host and certain regulatory proteins derived from the bacterium play important roles.

Knowledge about immunity control in λ accumulated from many years of research in numerous laboratories, progressing from genetic analysis to studies of protein chemistry and DNA sequence. Mutants of λ that are cI$^-$, cII$^-$, or cIII$^-$ cannot form lysogens but do grow lytically, thus giving clear plaques on the indicator bacteria. However, these mutants will not grow in a λ lysogen. Mutants that are $o_R^-o_L^-$ (called vir) also cannot form lysogens; they grow vegetatively even in a λ lysogen. Mutants that are N$^-$ or Q$^-$ do not grow even in nonlysogens.

Integration of Temperate Phage into the Host Genome

In λ, the final step in lysogenization is integration of the phage DNA into the host chromosome. It might be recalled that the phage genome is replicated while repression is being established, but when repression is established, the genes for DNA replication (operon 2 controlling middle functions) are shut off and the λ genome ceases to multiply. One of these λ chromosomes is able to integrate into the host chromosome by a site-specific recombination between a particular sequence in the phage (PP' in Fig. 11.3) and a particular sequence in the host chromosome (BB', which lies at about 15 minutes on the E. $coli$ map). Efficient integration depends on the int gene product, which is an enzyme that binds directly to the PP' and BB' sequences and catalyzes the crossing-over reaction.

Synthesis of the int protein occurs during the lysogenic response

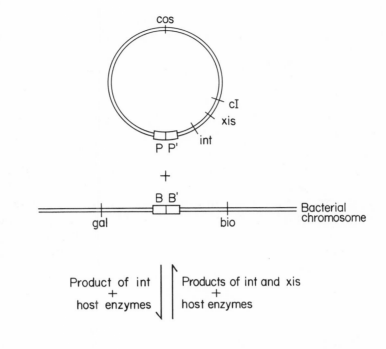

Fig. 11.3 Integration and excision of the phage genome. The circular phage chromosome undergoes recombination with the E. coli chromosome at a specific site designated PP' on λ and BB' on the host chromosome. The insertion of λ DNA into the host chromosome (integration) requires an enzyme encoded by the λ int gene as well as enzymes from the host. Reversal of this process (excision) leads to reconstitution of separate circular host and phage genomes. Excision requires the products of both int and xis genes of λ and enzymes from the host.

but not during the lytic response. Located in the middle of the xis (excision) gene on λ is a promoter, p_I (see Fig. 11.2), that is activated by the cII and cIII proteins, the same two proteins that establish repressor synthesis. The mRNA initiated from p_I contains the sequence coded by the int gene, and this mRNA is translated into the int protein. Even though the early transcript from p_L, propagated by the N protein, includes the int gene, it is not translated to give the int protein. This is because, unlike the transcript from p_I, the transcript from p_L continues beyond the *int* region to include the *sib* site. This site blocks translation of *int* mRNA, a phenomenon known as **retroregulation.**

Phage λ integration results in a new gene order that is an inverted permutation of the sequence of the original phage in linear form. Thus, the sequence of genes in the linear infecting phage is A-J-PP'-N-R, whereas the sequence in the lysogen is (bacterial genes)-BP'-N-R-A-J-PB'-(bacterial genes). The prophage can now be replicated as a part of the bacterial chromosome.

Not all temperate phages use the same mechanism for integration. In fact, it appears that the mechanism of integration of *E. coli* phage Mu is very similar to the insertion mechanism of transposons (see Chapter 8). P1, another lysogenic phage of *E. coli*, need not integrate at all; the prophage is usually maintained as a self-replicating plasmid complete with its own mechanisms for copy number control and incompatibility.

Induction of Lysogens

Lysogens are not highly stable. Repression is spontaneously lost at a frequency of about 10^{-3} in a population of cells, and lytic growth occurs following excision of the prophage. Induction of the latent phage can follow certain treatments such as irradiation by ultraviolet light or exposure to chemical mutagens. As a consequence of DNA damage by these agents, the cellular *recA* protease is activated and cleaves the λ repressor protein. This leads to a series of transcriptional events very similar to that in lytic multiplication after infection. The one important difference is that both the *int* and *xis* proteins must be synthesized in order to excise the prophage and restore the circular λ chromosome (Fig. 11.3). An apparent paradox arises here. Transcript 1 made from the infecting circular chromosome is not translated into the *int* protein, whereas transcript 1 made from the integrated prophage is. The reason for this difference is that the *sib* site lies to the left of the attachment site as drawn in Figure 11.1. Upon integration the *sib* site is disconnected from the rest of the DNA sequences of transcript 1, and can no longer participate in retroregulation.

Specialized Transduction

In Chapter 10 the curious ability of certain phage assembly systems to package headfuls of bacterial DNA was mentioned as the basis of generalized transduction. **Specialized transduction** occurs by a different mechanism. Here, the phages can transfer only those bacterial genes that are near the site of integration into the chromosome. Induction of a λ lysogen usually results in perfect excision and restoration of the circular λ chromosome (Fig. 11.3), but on very rare

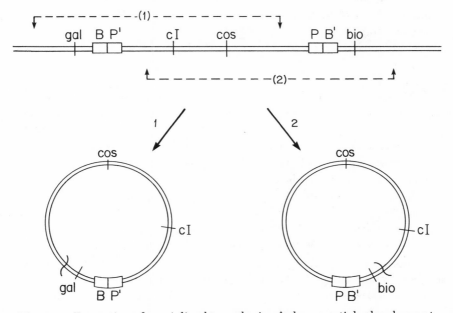

Fig. 11.4 Formation of specialized transducing λ phage particles by aberrant excision. A circular λ DNA molecule undergoes a recombination event with the host chromosome at a particular site on the phage DNA (called here *att* POP′) and a particular site on the bacterial DNA (called here *att* BOB′). The recombination event linearly inserts the λ DNA into the host chromosome. Reversal of this recombination ("excision") leads to reconstitution of the host chromosome and of the circular phage genome (att BOP′ $\times$ att POB′). If, however, the excision mechanism does not operate properly, bacterial genes adjacent to the integrated phage genome may be excised with the phage DNA, while some phage genes may be left behind. This leads to the formation of transducing phage, carrying nearby bacterial genes, such as *gal* and *bio*, respectively responsible for galactose utilization and biotin synthesis.

occasions the crossover does not occur through BP′ and PB′ but rather through other sites (Fig 11.4). In many cases these excised circles can go through lytic growth. Thus rare mature particles carrying bacterial genes close to the site of prophage insertion are present in the lysate (at a frequency of roughly 10^{-3} of cell populations). These special phages, carrying either the *gal* or *bio* bacterial gene clusters, may be recovered in the form of lysogens by infecting bacterial host cells with wild-type phages under appropriate selective conditions. For example, one could find λ phages carrying the gal^+ gene by selecting for lysogens of galactose-negative bacterial recipients that have acquired the ability to grow on this sugar. Acquisition of this growth property reflects integration of the specialized phage genome at the bacterial *att* site, making the cell heterozygous for *gal* (a copy of gal^- in normal position and a copy of gal^+ on the prophage).

Depending on the nature of excision, the *gal*-carrying λ may or may not be missing essential phage genes; λ packaging depends on specific cutting at the *cos* sites (cohesive ends); therefore the size of the packaged genome can be occasionally somewhat larger or smaller than normal. If the excised segment of DNA is missing certain phage genes, it is said to be defective. Certain imperfect phage genomes may be able to integrate, but induction of the defective lysogen will not give a lysate. However, the function of the missing genes can be supplied by a co-integrated normal λ phage; induction would yield both kinds of phages in about equal numbers. In such a lysate, instead of about 1 in 10^6 phages being gal^+, half would be. Aberrant excision of λ is analogous to the formation of F′ plasmids from the Hfr chromosome, except that a strict upper limit is imposed on the size of a recoverable phage genome.

Lysogenic Conversion

Resistance to superinfection by the *same* kind of phage as a result of phage repressor synthesis is not the only phenotypic change conferred by lysogenization. Some prophages alter the cell surface of their hosts in such a way that resistance to a variety of different phages is conferred. This phenomenon has been studied in particular detail for the "conversion" (change) of *Salmonella* membrane O-antigens by phages. In the example shown in Figure 11.5, an uninfected

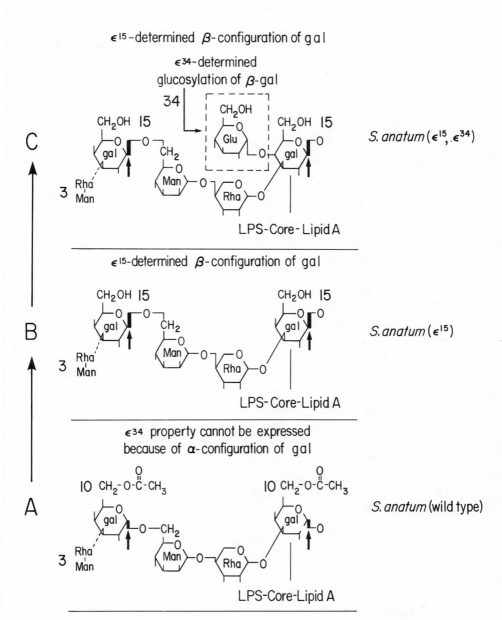

Fig. 11.5 Lysogenic conversion of O-antigen in *Salmonella* (see text).

strain of *Salmonella* has a particular O-antigen whose determinants, 3 and 10, are respectively the mannosyl-rhamnose disaccharide and the acetyl group on a galactose unit. After this strain is infected with phage ϵ^{15}, the O-antigen changes chemically, losing determinant 10 and gaining determinant 15 by rearrangement of the determinant-bearing galactose (from galactosyl-α-1, 6-mannose to galactosyl-β-1,6-mannose). The galactose acetylation reaction by the cell is stopped, possibly by the production of a repressor for the bacterial gene encoding the acetylase. In addition, an inhibitor for the step making the α-galactosyl-mannose linkage (this is the trisaccharide polymerization step in O-antigen synthesis) and an enzyme that makes β-galactosyl-mannose both appear. The old antigen is thus lost, and the new one is no longer a good receptor for phage ϵ^{15}. However, it happens to be the receptor for a different phage, ϵ^{34}, whose entrance and establishment in turn causes another modification of the O-antigen (a new residue, glucose, is joined to the galactose). Thus the nature of these particular O-antigenic determinants seems to depend on the resident temperate phages.

Lysogenic conversions can be very important to the course of bacterial infections. The loss of one O antigen and its replacement with another may have, depending upon which antibodies are present in the host, tremendous survival value for the bacterium and may prolong the length of an infection. This would be analogous to phase variation as outlined in Chapter 5.

There are also other cases of phage-induced immunity in which the lysogen seems to have altered its envelope in such a way that the phage can still attach but DNA injection is prevented. Yet another mechanism is seen with phage λ. This phage carries two genes next to cI, called rexA and B, whose products are expressed in lysogens. These proteins provide λ lysogens with a degree of resistance to a number of completely unrelated phages.

In the above examples, the lysogenic conversions are evidently related to protection against superinfection. There are now numerous examples of lysogenic conversions affecting functions that bear no direct relationship to development of the phage itself. There are phages that carry antibiotic resistance genes (sometimes transposons), phages that carry particular restriction-modification systems (see Chapter 12), and — most important in medical bacteriology — phages that carry genes for virulence factors of pathogens. All of these phage-determined traits of the host seem to reinforce the notion that temperate viruses and their bacteria have established a mutualism.

The ability of Corynebacterium diphtheriae to cause disease is dependent on a gene carried by a temperate prophage called β. This gene, tox, codes for the diphtheria toxin. Normally the tox gene is repressed in diphtheria bacteria lysogenic for β; this is caused by an Fe^{++}-dependent repressor synthesized by the bacterium. In an environment deficient in free Fe^{++} (the human body, for example), the tox gene is expressed. Excretion of this toxic protein causes necrosis of neighboring tissue cells in the human host (usually beginning in the upper respiratory tract). The cellular material released stimulates further growth of the pathogen.

Questions

11.1. How does the formation of F′ plasmid in an Hfr bacterium resemble the formation of specialized transducing phages, such as λ gal?

11.2. Certain mutations in phage β prevent toxin production by Corynebacterium diphtheriae without altering lysogeny. (a) Why does the above information not suffice to prove that the toxin structural gene is actually carried by the phage? (b) How could one prove that the gene is carried by the phage?

11.3. Temperate phage $\phi80$ can transduce several chromosomal genes, including A, B, and C, after induction of a lysogen. Transduction frequencies are: A, 10^{-7} transducing particles/phage; B, 10^{-7} transducing particles/phage; and C, 5×10^{-9} transducing particles/phage. Moreover, $\phi80$-A particles rarely also carry C and never B; $\phi80$-B particles never carry C or A; $\phi80$-C particles always also carry A, but never B. What is the order of the three genes and the site of $\phi80$ insertion?

11.4. Group B Streptococcus is an important pathogen which is generally sensitive to penicillin G. However, a penicillin-resistant strain was found from a clinical specimen. Sensitive cells appear spontaneously in populations of this strain at a frequency of about 10^{-3}. The resistance is due to the synthesis of a penicillinase indistinguishable from the enzyme isolated from Staphylococcus. Both sensitive and resistant group B Streptococci carry different plasmids, but the resistant bacteria carry a plasmid which contains a seven-kilobase sequence not found in any of the plasmids in sensitive strains. DNA hybridization shows that plasmids from Streptococci and Staphylococci have no homology except this seven-kilobase sequence which is associated with the drug resistance. No Staphylococcus plasmid has been successfully introduced and propagated in any of group B Streptococcus under experimental conditions.

The penicillinase in the resistant strain of Streptococcus is synthesized at an increased rate within a few minutes after the cells are exposed to the drug. If the cells are exposed to ultraviolet light, lysis occurs within two hours. When an appropriately diluted culture filtrate is added to lawns of certain other group B Streptococci, turbid plaques are formed. Only about 10^{-6} of those plaques contained colonies that are penicillin resistant. However, all of the bacterial colonies from such a rare plaque are drug resistant.

(a) What might be the nature and function of the seven kilobase sequence? (b) What might be the origin of this sequence in resistant strains of Streptococci? (c) What is the likely mechanism underlying the increase in the synthesis of penicillinase by the resistant Streptococcus shortly after exposure to the drug? (d) What is the nature of the rare (10^{-6}) plaques in which all Streptococci lysogens turned out to be drug resistant?

Further Reading

Campbell, A. 1981. Evolutionary significance of accessory DNA elements in bacteria. Annual Review of Microbiology 35:55–83.

Echols, H., and H. Murialdo. 1978. Genetic map of bacteriophage λ. Microbiological Reviews 42:577–591.

Eisen, H., P. Brachet, L. Pereira da Silva, and F. Jacob. 1970. Regulation of repressor expression in λ. Proceedings of the National Academy of Sciences, USA 66:855–862.

Greenblatt, J. 1981. Regulation of transcription termination by the N gene protein of bacteriophage λ. Cell 24:8–9.

Hershey, A. D., ed. 1971. The Bacteriophage λ. New York: Cold Spring Harbor Laboratory.

Herskowitz, I, and D. Hagen. 1980. The lysis-lysogeny decision of phage λ: explicit programming and responsiveness. Annual Review of Genetics 14:399–445.

Hochschild, A., N. Irwin, and M. Ptashne. 1983. Repressor structure and the mechanism of positive control. Cell 32:319–325.

Meyer, B., D. Reid, and M. Ptashne. 1975. λ repressor turns off transcription of its own gene. *Proceedings of the National Academy of Sciences, USA* 72:4785–4789.

Schindler, D., and H. Echols. 1981. Retroregulation of the *int* gene of bacteriophage λ: control of translation completion. *Proceedings of the National Academy of Sciences, USA* 78:4475–4479.

Weisberg, R., S. Gottesman, and M. Gottesman. 1977. Bacteriophage λ: the lysogenic pathway. *Comprehensive Virology* 8:197–258.

12

DNA Restriction and Gene Cloning

GENETIC ELEMENTS (such as the F plasmid and phage λ), which are capable of self-replication and can incorporate bacterial genes as stable parts of their genomes, have proven to be important in the basic research into many processes of molecular genetics. A major reason is that once a bacterial genetic unit has been incorporated into one of these DNA **vectors,** it becomes possible to purify large amounts of a desired DNA sequence. Rigorous biochemical studies of the purified bacterial gene can then be performed. Studies on specialized transducing phages carrying bacterial genes contributed importantly to the elucidation of the regulatory mechanisms given in Chapter 5 — negative control of the lactose operon, positive control of the arabinose operon, and the attenuator control of the histidine operon.

Recombinant DNA technology has now been extended to a point where it is feasible to purify single genes not just from bacteria but from *all* organisms. This development was made possible by understanding the molecular genetics of E. coli and its interactions with infecting phages or other foreign DNA. Much of this development was described in earlier chapters. However, before embarking on a description of recombinant DNA technology, it is necessary to introduce the phenomenon of restriction and modification.

Restriction and Modification

Bacteria can defend themselves against infecting phages (or other generally harmful genetic elements) by attacking the intruding DNA at specific vulnerable sites in the molecule. This primitive immune system is called **restriction,** and the vulnerable base sequences are called **restriction sites.** Though phages normally have broad host ranges (that is, a given species of phage can attach to and inject its DNA into many different strains or even different species of bacteria), infection often is kept in check by restriction, if a phage grown on one bacterial strain is introduced into a second strain. For example, if phage P2 is grown in $E.$ $coli$ strain C and then used to infect strain K, the plating efficiency is very low (about 10^{-3}); in contrast, if the same phage preparation is used to infect $E.$ $coli$ strain C, the plating efficiency is close to one. Strain C is lysed efficiently because it does not have a restriction system for protection against the infecting phage P2 DNA. In the case of strain K, on the other hand, the phage adsorbs to the cells and injects its DNA in the usual way, but the DNA is rapidly degraded by a specific endonuclease that cuts the double-stranded DNA, thereby terminating the infection. A nucleolytic enzyme capable of cutting DNA in this manner is called a **restriction endonuclease.**

This type of bacterial immunity immediately raises a number of questions: for example, can any foreign DNA escape cleavage by restriction enzymes, and how is the cell's own DNA protected from its restriction endonuclease? A clue to the answers was provided when the small fraction of P2 particles that had successfully replicated in $E.$ $coli$ strain K (as described above) was again used to infect cells of strain K: the plating efficiency was high and essentially all of the emerging phage particles were competent to infect and multiply in strain K. These phages had seemingly been modified in some way during growth inside the K cells so that they acquired immunity to the K restriction endonucleases. It was eventually learned that the host cell DNA (which carries vulnerable restriction sites like those in P2 DNA) is protected from restriction in the same way.

The process, called **modification,** involves methylation of certain bases within the vulnerable site (a seven base-pair sequence), thereby protecting the DNA from the restriction endonuclease. In the DNA of $E.$ $coli$ strain K, certain adenines within the specific seven base-pair target site are methylated at the N-6 position by an enzyme with S-adenosylmethionine as the methyl donor. The bacterial DNA contains the same N-6 methyladenines wherever these same seven base-pair sequences appear. The cell thus safeguards its DNA from degradation by modifying its restriction sites. A cell which carries a gene for a restriction endonuclease must necessarily also carry a gene for an appropriate modification methylase. This set of enzymes in *E. coli* K is referred to as the **K system** of restriction and modification.

When methylated DNA (whether viral or chromosomal) replicates, the newly synthesized DNA is modified almost immediately (within two minutes in *E. coli* DNA). If phages with K-modified DNA are again grown on *E. coli* C, all of the newly synthesized DNA strands will be nonmethylated at the special adenine residues, since strain C has no restriction-modification system of its own. Thus the emerging phages will no longer have a high plating efficiency on strain K, although their ability to grow on strain C is unchanged. Figure 12.1 shows growth patterns of phages grown alternately on strains K and C.

Genes encoding a restriction-modification system can be found on the bacterial chromosome (as in the K system of *E. coli*), on plasmids (as in the EcoRI system, so named because it is found on the plasmid EcoRI), or on phages (as in the P1 system). These different restriction-modification systems act on different specific DNA sites. Thus, an *E. coli* K strain harboring both the EcoR1 plasmid and lysogenic phage P1 would have at least three different restriction-modification systems which could act on an entering genome of phage P2 grown in *E. coli* strain C. The presence of restriction-modification systems in plasmids and phages suggests competition among these genetic elements. The element that gets established first not only has a chance to monopolize the cell but also compensates the host by providing protection against other infections, thereby maintaining a symbiotic relationship.

The sensitivity of a particular DNA to a particular restriction system depends partly on the frequency of its vulnerable restriction sites. The specific sequence of bases that make up a restriction site is long enough to be statistically rare; even in a large phage such as λ there are only four sites for K restriction. Phage T7, whose genome is only slightly smaller, has none. Hence T7 grown in any host is able to infect strain K with impunity. The rarity of certain restriction targets might be the result of natural selection.

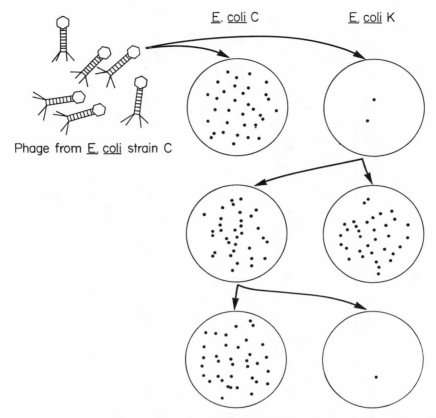

Phage from E. coli strain C

Fig. 12.1 Host restriction of phage growth. Phage particles grown in *E. coli* strain C are used to infect lawns of *E. coli* strain C or strain K.

BIOCHEMISTRY OF RESTRICTION AND MODIFICATION

On the basis of known DNases of different specificity that behave like restriction endonucleases, there should be well in excess of 70 different restriction-modification systems in the prokaryotic world. These fall into three general categories.

Type I In these systems (of which the K system is one), the endonuclease recognizes unprotected restriction sites but then cuts the DNA molecule at some variable distance away from them. This means that the DNA fragments cut out of a population of identical DNA molecules by cleavage at two given restriction sites will differ both in size and in terminal sequences. The restriction enzyme

Table 12.1. Recognition sequences (restriction sites) for a few representative restriction endonucleases.

Type	Example enzyme	Recognition sequence[a]	Digestion products
I	K	$\overset{*}{A}AC(N)_6GTGC(N)_n$ $TTG(N)_6CACG(N')_n$ $\underset{*}{}$	Fragments produced from identical molecules will be of variable size. The terminal sequences of fragments between two given restriction sites will differ from molecule to molecule and are typically more than 1 kb from the recognition site.
II	EcoRI	$G \overset{\downarrow}{} A\overset{*}{A}TT\ C$ $C\ TTAA \underset{*}{\uparrow} G$	Fragments produced from identical molecules will fall into discrete-size classes, the number of which is determined by the number of recognition sites in the original molecule. From class to class, the termini are unique; they may be single-stranded for an enzyme that produces staggered cuts (like EcoRI) or double-stranded for an enzyme that cuts directly across (like Bal I). See Figure 12.2.
	Bal I	$TGG \overset{\downarrow}{} CCA$ $ACC \underset{\uparrow}{} GGT$	
III	P1	$\overset{*}{A}GACC(N)_{24} \overset{\downarrow}{} N\ N\ N$ $TCTGG(N')_{24}\ N'N'N'$	Fragments produced from identical molecules will fall into discrete-size classes, as in the type II system. However, the terminal sequences on different fragments within a class will differ, although they are at a constant distance from the recognition site.

a. The arrow shows the cleavage point. Asterisk indicates bases that are methylated during modification. N represents any base, and N' its complementary base.

requires both the hydrolysis of ATP and the presence of S-adenosyl-methionine to cut susceptible DNA.

Type II Endonucleases of Type II cut within the restriction site itself. Between two given restriction sites the digestion product will always be the same size from molecule to molecule, as is shown in Figure 12.2. The base sequences at the ends of fragments will be those found within the restriction site. These ends may be single-stranded or double-stranded, depending on the particular enzyme.

Type III This class of endonucleases recognizes its restriction site and then cuts the DNA molecule a fixed number of bases away. Thus the fragments cut out between two given sites will be of the same size from molecule, but the DNA sequences at the ends of those fragments will probably be different.

The type of restriction-modification system cannot be predicted on the basis of the kind of genome (plasmid, phage, or bacterial chromosome) with which it is associated. Table 12.1 shows the nucleotide sequence of representative examples of the three types of restriction sites and how they are cut. Note that with type II enzymes, two different kinds of cuts are shown; the EcoRI enzyme staggers cuts on opposite strands to give single-stranded tails on the digestion products, whereas the Bal enzyme (so called because it is isolated from *Brevibacterium albidum*) cuts directly across both strands to produce double-stranded blunt ends.

Formation of DNA Hybrids

It is type II restriction endonucleases that are important in the manipulation of isolated genes in recombinant DNA research, because between two given restriction sites they produce fragments whose size and termini are uniform from molecule to molecule. After a large DNA molecule is digested by such an enzyme, a given base sequence of interest will likely be contained in one of the discrete fragments. For example, in Figure 12.2 phage λ DNA molecules have been cut into fragments of five sizes, which can then be resolved by their different migration rates during electrophoresis in a

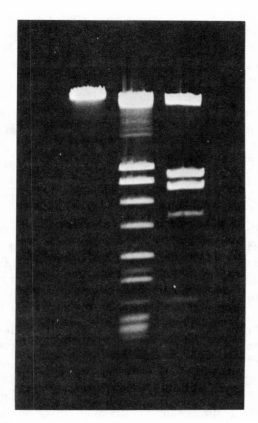

Fig. 12.2 Electrophoresis patterns of λ DNA after digestion with two restriction enzymes. DNA molecules can be separated according to their size by electrophoresis through agarose gels. In this procedure DNA solutions are put in narrow rectangular clefts along one border of an agarose sheet placed flat in the apparatus (top side of the photograph). An electric field is then applied across the gel. DNA, which is highly negatively charged, will move toward the anode (the "+" pole). The agarose gel retards DNA movement, permitting the smaller molecules to move faster than the larger ones. The migration rates are inversely proportional to the logarithms of the molecular weights of the DNA. After electrophoresis the DNA is stained with the fluorescent dye, ethidium bromide, activated with ultraviolet light, and photographed with visible light. (As little as $0.05\,\mu g$ of DNA can be detected.)

In the figure three DNA samples were placed in separate wells. The lane on the left contains intact λ DNA and the band shows the migrated position of the DNA at the end of the experiment. The center lane contains λ DNA cut with the enzyme Bal I, and the right lane contains λ DNA cut with the enzyme EcoRI. The formation of new DNA fragments of discrete size after digestion shows that these two enzymes cut at specific sites. This procedure is widely used for determining the number of cutting sites in a DNA molecule, as well as the distances between the sites.

gel. More importantly, each group of separated DNA fragments of a given size can be recovered and inserted into copies of a plasmid or a viral genome of choice. The newly constructed vector can then be propagated autonomously in a prokaryotic or eukaryotic cell. How the *lacY* gene of E. *coli* could be purified using these techniques is illustrated in Figure 12.3 and described below.

GENE PURIFICATION

The E. *coli* chromosome is 4,000 kb long, of which 1.3 kb is the *lacY* gene. In order to purify this gene, the first step is to cut copies of the bacterial chromosome into fragments with the EcoRI restriction enzyme. The six base-pair sequence recognized by EcoRI endonuclease would be expected to appear about once every 4 kb if the DNA sequence were random, so about 1,000 fragments per E. *coli* chromosome would be anticipated. Unless the enzyme cleaves *lacY* itself, one of these fragments will carry the gene, which must be eventually singled out. Statistically, there is about 0.3 probability that a gene of this size will carry the six base-pair sequence recognized by a restriction enzyme, resulting in a cut within the gene. In that case a restriction enzyme recognizing a different sequence can be used.

The next step requires purification of a suitable vector into which the fragments can be inserted. A derivative of the ColE1 plasmid called pBR322 (4.4 kb) would be appropriate in this case because it carries a β-lactamase gene (amp^R), which can be used later in the selection procedure, and because it has a single EcoRI site. Digestion of the pBR322 DNA with the EcoRI enzyme therefore converts the plasmids from circular to linear molecules with short single-stranded 5'-AATT ends. If linear DNA from such plasmids is mixed with the cut E. *coli* fragments (which also have 5'-AATT ends), some of the plasmid DNA will anneal with bacterial DNA under favorable conditions. The mixture is next treated with DNA ligase, an enzyme that covalently joins nicked single strands by formation of phosphodiester bonds. The ligation reaction is analogous to the covalent closing of phage λ DNA once it has been inserted into a cell.

Among the products will be circular plasmid molecules with inserted bacterial DNA, but there will also be plasmids joined to plasmids and bacterial fragments joined to other bacterial fragments. However, if an appropriate concentration of DNA is used in the ligation reaction, only moderately sized circles composed of no more

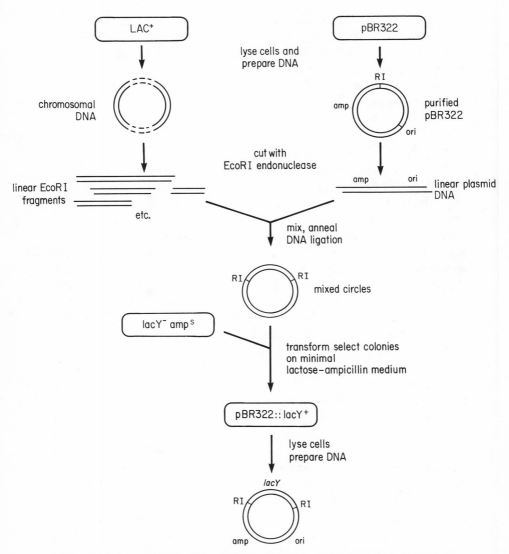

Fig. 12.3 Insertion of a segment of *E. coli* DNA containing the *lac* genes into the plasmid pBR322.

than two or three fragments will be the principal products. Even then only a tiny fraction of the circular molecules formed will be pBR322 plasmids containing the *lacY* gene as an insert. This particular hybrid can be fished out by adding the mixture of circular molecules to bacteria that are amp^S and $lacY^-$ and then growing them on lactose medium to which ampicillin has been added. Only those transformants that have acquired ampicillin resistance *and* the ability to use lactose from the pBR322 plasmid will grow. This strain will now carry a new plasmid composed of pBR322 and the $lacY$ sequence.

The $amp^R,lacY^+$ bacterial clone is then grown in liquid culture and the plasmid DNA purified free from the contaminating bacterial DNA. The $lacY^+$ gene will be in a plasmid of about 8 kb (as opposed to the 4,000 kb chromosome where it started out) and therefore is enriched about 500-fold. This gene can now be easily recovered by cutting the plasmid DNA with different restriction endonucleases and isolating the appropriate fragment by gel electrophoresis. Clearly, such a protocol dictates that the enzyme used to clone a gene should cut neither the selected genes (amp^R and $lacY^+$ in this case) nor genes essential to the replication of the plasmid. Given that there are many different restriction endonucleases and many different cloning plasmids, finding a suitable combination usually is not difficult.

In the above example, clones that carried the $lacY^+$ gene were selected in one step by requiring the transformant to grow on lactose in the presence of ampicillin. Alternatively, a few thousand individual amp^R clones can be selected and these can be subsequently screened individually for the desired trait. This more cumbersome procedure has to be used when the trait offers the transformed bacterial cell no selective advantage (such as the ability to grow on a special carbon source). Techniques by which individual bacterial colonies can be screened for the presence of a specific protein or DNA sequence include rendering a portion of the cells in the bacterial colony accessible to specific antibodies against the protein or to a short ^{32}P-labeled DNA sequence that will anneal with the cloned gene by homology. Screening by DNA homology has the advantage of not requiring phenotypic manifestation of the cloned gene.

Phage λ is often used to clone DNA fragments. Special derivatives of the phage that carry a single EcoRI site, or two EcoRI sites flanking a segment of phage DNA not essential for reproduction in E. *coli*, have been constructed. Foreign DNA of appropriate lengths can be inserted into the single EcoRI site or used to replace the nonessential λ DNA flanked by the two EcoRI sites with methods similar to that used for inserting DNA fragments into the plasmid pBR322. Individual phages with foreign DNA inserts are usually propagated by lytic growth in bacterial lawns, each plaque representing a different clone. The protein product or desired DNA sequence may be screened in the plaque. This is possible because a single plaque is also a highly concentrated solution of phage-specified proteins and nucleic acids.

Interspecies Hybrids

In principle, any linear piece of DNA with a 5′-AATT sequence on each end of the fragment can be inserted into an EcoRI site of pBR322 and then cloned (amplified) in $E. coli$. Indeed, even linear DNA fragments from a plant or animal source will serve; there are no species barriers that prevent alien fragments from being replicated in $E. coli$, as long as they are joined to an acceptable ori site. This fundamental finding has given birth to the gene cloning technology that is revolutionizing genetic research.

Replication of a cloned gene, however, does not guarantee its expression. In fact genes from higher organisms are rarely expressed in the bacterial cell. This is probably due to the inability of the transcription and translation system of $E. coli$ to recognize the foreign promoters and translational initiation signals. More importantly, the vast majority of eukaryotic gene transcripts require processing that involves splicing. In addition, some polypeptides might not assume their proper form in a foreign cytoplasm. Even if correctly formed, they might be susceptible to rapid proteolytic degradation. Until these problems are solved, the yield of foreign proteins will be low or nonexistent. But with appropriate manipulations, these barriers against expression have been overcome in a few cases.

CLONING OF A MAMMALIAN β-GLOBIN GENE IN E. COLI

An example of how a eukaryotic gene may be cloned into $E. coli$ and subsequently analyzed and manipulated to direct the synthesis of a protein is provided by recombinant studies of the β-globin gene. The DNA for a single mammalian gene will be an infinitesimal fraction of total cellular DNA. For example, a gene of 3 kb in a typical mammalian genome of 3×10^6 kb comprises only 0.0001% of total cellular DNA. When digested with EcoRI, the rabbit genome is cut into approximately 10^5 fragments (much of the rabbit genome consists of

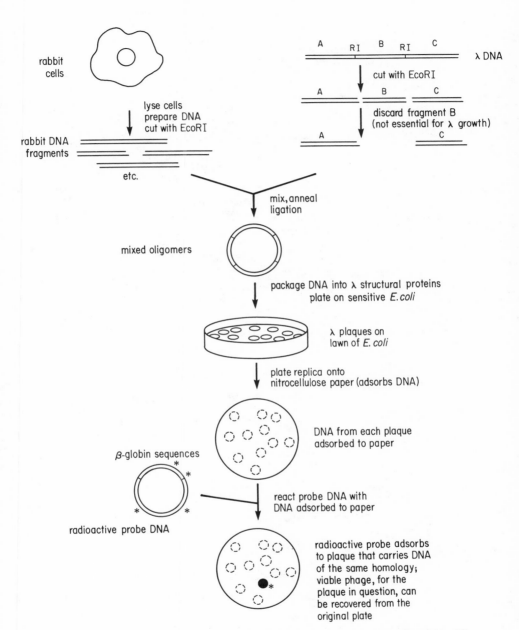

Fig. 12.4(A) Cloning the rabbit β-globin gene into a λ vector and its identification by screening with a probe.

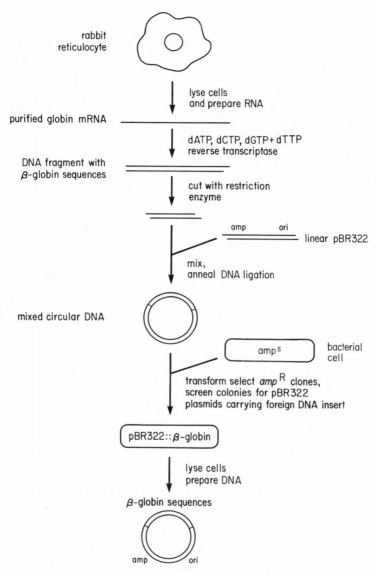

Fig. 12.4(B) Preparation of a probe for the β-globin gene.

repetitive noncoding sequences), only one of which carries the β-globin gene. This collection of EcoRI-produced fragments is passed through a crude purification step to enrich the desired fragment (as revealed by radioactively labeled homologous sequences; see below) and then mixed with EcoRI-digested λ DNA (Fig. 12.4A). Conditions are used so that on the average one rabbit DNA fragment is inserted into each λ chromosome. To ensure that almost all λ chromosomes carry a foreign insert, an undersized λ DNA chromosome can be tailored. Unless this small chromosome is properly lengthened by an insert, the packaging mechanism will not produce an infective particle. Infective particles that contain the special λ chromosome carrying a foreign DNA insert are plated out in bacterial lawns to obtain single plaques. One plaque in 10,000 to 30,000 should carry the β-globin gene.

The next problem to be overcome is the lack of a phenotypic signal to indicate which plaque carries the desired gene. One way to develop a screening assay is to purify the β-globin mRNA, since it can provide the unique base sequence that will identify the β-globin gene. Rabbit reticulocytes are rich in β-globin mRNA and therefore a good source for its purification (Fig. 12.4B). Once purified, mRNA molecules can be used as a template for the synthesis of double-stranded DNA. Polynucleotide polymerases called reverse transcriptases can copy the purified β-globin mRNA into DNA. These synthetic DNA molecules can be cut with restriction enzymes and manipulated like the DNA isolated from chromosomes. Thus the synthetic DNA with rabbit β-globin sequences can be cloned into a plasmid such as pBR322. This DNA is labeled with ^{32}P so that it may be used as a **probe** for homologous base sequences in the rabbit chromosomal DNA. For the purpose of identification the base sequence need not encompass the entire gene that is sought.

Screening is a multi-step process which involves hybridizing the DNA in each plaque with the radioactive DNA copied from the cloned β-globin sequence. The plaque that contains the β-globin gene should have DNA sequences that form duplexes with the radioactive probe. The clone of λ thus identified will carry the entire β-globin gene. This phage is grown in large volume for the preparation of quantities of DNA rich in the β-globin gene.

The β-globin genes from a variety of sources, including human, have been cloned by using procedures similar to the one described here. Highly enriched fragments containing the β-globin sequences were used to determine the fine structure and complete nucleotide sequence of a mammalian gene for the first time. This information led immediately to a molecular understanding of a variety of β-thalassemias (inherited β-globin deficiency). In addition, knowledge of the fine structures of this gene and of others like it have laid a foundation for studying gene expression in higher organisms.

EXPRESSION OF β-GLOBIN IN BACTERIA

$E.\ coli$ infected with λ bearing the β-globin gene does not produce the β-globin protein, for three reasons. First, the rabbit β-globin promoter is not recognized by the RNA polymerase of $E.\ coli$. But even if the gene were transcribed (for example, by means of one of λ's promoters), the resulting RNA would not be translated, since $E.\ coli$ ribosomes do not bind to the β-globin transcript. Finally, even if the ribosomes could bind to the β-globin transcript, the β-globin protein would still not be synthesized because there are sequences inside the structural gene for β-globin (called introns) that are transcribed into RNA but must be removed before these transcripts become mRNA. This removal (called intron processing) is carried out by a system in rabbit cells not present in $E.\ coli$.

In order to construct a clone that synthesizes β-globin, the following manipulations are necessary. The complete DNA sequence made from mature β-globin mRNA, which has no introns, is cloned into a plasmid, as in Figure 12.4A. However, the sequence has the rabbit ribosome binding site in front of the initiation codon. With the help of appropriate enzymes, this site is deleted and replaced by a bacterial lac promoter and the ribosome binding site for the $lacZ$ gene. When the reconstructed plasmid is transformed back into $E.\ coli$, the bacteria will synthesize β-globin protein, but only when lactose or another inducer is added to the growth medium. This shows that the lac promoter is directing the expression of β-globin protein.

With the application of procedures such as these, numerous proteins from eukaryotic sources have been expressed in bacteria. Those of potential value for human health include human insulin, growth hormones, interferon, and viral antigens for use in vaccine production.

Questions

12.1. Suggest a mechanism for safeguarding newly synthesized host DNA from restriction.

12.2. There are three strains of $E.\ coli$: A, B, and C; and three strains of phage: ϕwt (wild-type) and its two mutants ϕX^-, and ϕY^-. Strain B differs from strain A in being lysogenic for phage ϕwt. When the phage strains are tested on the bacterial strains, the following results are observed:

Phage	E. coli A	E. coli B	E. coli C
ϕwt	turbid	no plaque	very rare (turbid)
ϕX^-	clear	no plaque	very rare (clear)
ϕY^-	clear	clear	very rare (clear)

(a) Explain phage mutants ϕX^- and ϕY^-, and speculate as to how one bacterial strain may differ from the others. (b) Strain C yields only 10^{-4} as many plaques as strain A. Give possible explanations of the rare plaques, and how the explanations could be tested.

12.3 There are three bacterial strains, identical except for mutational diferences in the restriction-modification system. When phage λ is grown on each of these three strains and the phage preparation from each is titered against the three strains, the results are as follows:

	Plaque forming units per ml when plated on:		
Phage λ pregrown on	Strain 1	Strain 2	Strain 3
Strain 1	3.8×10^8	4.1×10^8	4×10^8
Strain 2	4.2×10^8	3.9×10^8	3.7×10^8
Strain 3	4.0×10^8	5×10^4	4.1×10^8

Give the genotype for each of the three strains. (+) for activity present; (−) for activity absent.

Strain	Restriction	Modification
1		
2		
3		

12.4. A plasmid which carries the transposon Tnl (conferring constitutive ampicillin resistance) and whose DNA is not restrictable in E. coli has been constructed (Fig. Q12.4). A preparation of this plasmid is treated with the restriction enzyme BamI and the resulting two plasmid fragments of unequal length, "big Bam" and "little Bam," are separated. A transformation experiment is performed with little Bam or big Bam, using as recipient a $lacR^-,Z^-,Y^-$, ampicillin-sensitive E. coli which is recA (defective in homology-dependent recombination). In

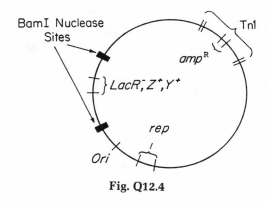

Fig. Q12.4

both experiments, a large quantity of DNA is used to ensure the infection of most of the cells with at least one fragment. The cells are incubated in a growth medium. One hour later, β-galactosidase and ampicillinase activities in the cell extract are assayed. At the same time, samples of the infected cells are spread on two different agar plates, one containing lactose as the carbon and energy source and another containing ampicillin with glucose as the carbon and energy source, and are checked for colonies after 24 hours. Fill in the following table to show positive or negative results from each of these four tests.

DNA fragment used to infect the cells	β-galacto-sidase activity	Ampicil-linase activity	Growth on agar containing	
			Lactose	Ampicillin + glucose
Little Bam				
Big Bam				

12.5. Suppose that with restriction enzymes and appropriate techniques, the gene for botulism toxin from a *Clostridium* species is connected to the DNA of a simple *E. coli* plasmid (containing a single site for the EcoRI restriction enzyme and a tetracycline-resistance gene not containing the restriction site). However, none of the *E. coli* transformed by such plasmids produces botulism toxin. What might account for the failure?

12.6. Phage λ DNA can be isolated either as linear or circular molecules. If λ has five EcoRI cutting sites, in what form was the λ DNA used in the experiment shown in Figure 12.2?

12.7. When penicillin was discovered as a natural product of the mold *Penicillium notatum*, the amount produced was small (for example, 5 μg/ml of culture), but nowadays for commercial production variants are used which produce large amounts (5 mg/ml). (a) How might the present strains differ from the original? (b) How might the present strains be further improved?

Further Reading

Arber, W. 1979. Promotion and limitation of genetic exchange. *Science* 205:361–365.

Bunn, H. E., and B. G. Forget. 1984. *Hemoglobin: Structure, Function, Molecular Genetics, Clinical Aspects.* Philadelphia: Saunders.

Clarke, L., and J. Carbon. 1976. A colony bank containing synthetic Col EI hybrid plasmids representative of the entire *E. coli* genome. *Cell* 9:91–99.

Cohen, S. N. 1980. The transplantation and manipulation of genes in microorganisms. *Harvey Lecture Series, 74.* New York: Academic, pp. 173–207.

Konkel, D. A., J. V. Maizel, Jr., and P. Leder. 1979. The evolution and sequence comparison of two recently diverged mouse chromosomal β-globin genes. *Cell* 18:865–873.

Kruger, D. H., and T. A. Bickle. 1983. Bacteriophage survival: multiple mechanisms for avoiding the deoxyribonucleic acid restriction systems of their hosts. *Microbiological Reviews* 47:345–360.

Maniatis, T., E. Fritsch, and J. Sambrook. 1982. *Molecular Cloning.* New York: Cold Spring Harbor Laboratory.

Old, R. W., and S. B. Primrose. 1981. *Principles of Gene Manipulation.* 2nd ed. Berkeley: University of California Press.

Smith, H. O. 1979. Nucleotide sequence specificity of restriction endonuclease. *Science* 205:455–462.

Yuan, R. 1981. Structure and mechanism of multifunctional restriction endonucleases. *Annual Review of Biochemistry* 50:285–315.

13

Chemotherapy and Antibiotics

ALTHOUGH most bacteria are not pathogenic, many are. Thus it is not surprising that much effort has been expended in search of drugs that can control disease-causing organisms. An added benefit of studies on the mechanism of action of many drugs has been important insights on distinctive features of the biochemistry and physiology of nonpathogens as well.

The field of chemotherapy was developed in the 1890s by one man, Paul Ehrlich. Having worked on the selectivity of dyes in staining tissues, and on the exquisite selectivity of antibodies in protecting against specific infectious agents, he combined the two approaches in a systematic search for **chemotherapeutic agents:** chemicals toxic to pathogenic organisms at concentrations tolerated by the host (selective toxicity). Ehrlich had limited success with arsenicals for trypanosomiasis and syphilis, and after his death in 1915 the field was dormant until the discovery in 1936 of **sulfonamides,** effective against streptococcal and a few other infections. The range of chemotherapy was soon extended much further by the discovery of **antibiotics:** products of microbes themselves that are inhibitory to various other microbes. The discovery of the first antibiotic, **penicillin,** resulted from Alexander Fleming's chance observation in 1922 that a contaminating colony of the mold *Penicillium notatum* lysed nearby colonies of staphylococci. It was not until ten years later that Ernst Chain isolated the active product of the mold and Howard Florey showed its dramatic effectiveness in patients.

Since then thousands of antibiotics have been isolated from various soil organisms (mostly streptomycetes), largely by the pharmaceutical industry. Of these, perhaps fifty have proved selective enough to be useful clinically; many of the others are useful reagents in research because each blocks some definite step in metabolism.

Because some antibiotics have now been commercially synthesized and some chemotherapeutic agents are derived from antibiotics by artificial modification, the general term **antimicrobial therapy** is now widely used to refer to the clinical use of any chemicals—whether synthesized by microbes or in the laboratory—which are toxic to pathogenic organisms at concentrations tolerated by the host. This chapter will deal with clinically important aspects of antimicrobial actions, including drug resistance, and also with various molecular mechanisms of action.

Tests for the Effects of Drugs

A key characteristic of any antimicrobial agent is its **antimicrobial spectrum:** the range of organisms that are sensitive to the agent at concentrations attainable in the host. Drug sensitivity may be determined by serial dilution: inoculating the organism in liquid media containing various concentrations of a compound and observing whether or not overnight growth is prevented. A more convenient method is agar diffusion: an appropriate amount of drug (usually in a commercially available filter disk) is placed on a solid medium heavily seeded with bacteria on its surface. After an incubation period, a clear zone around the disk in the opaque lawn of bacterial growth indicates antibacterial activity. The diameter of the zone roughly indicates the degree of sensitivity (though the size of the zone around different agents is not the only basis for choice in therapy).

BACTERIOSTATIC AND BACTERICIDAL ACTION

Some antimicrobial agents are **bactericidal** (lethal to bacteria) while others are only **bacteriostatic** (they reversibly inhibit bacterial growth). Turbidimetric measurements of the kinetics of growth inhibition, following addition of a drug to a growing culture (Fig. 13.1), do not distinguish the two actions unless the killing is associated with cell lysis.

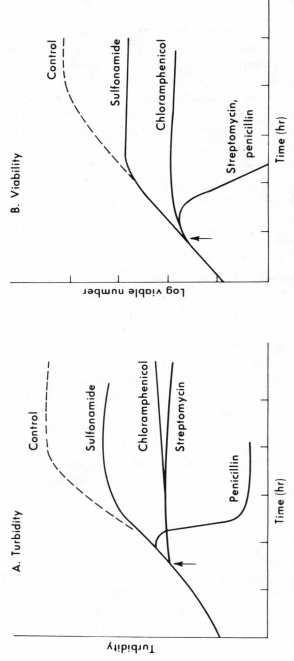

Fig. 13.1 Kinetics of the antimicrobial action of representative drugs. Drugs are added at the arrow to an exponentially growing culture. Penicillin is not effective if added to a culture in stationary phase. (From B. D. Davis, R. Dulbecco, H. N. Eisen, and H. S. Ginsberg. 1980. *Microbiology*. 3rd ed. New York: Harper and Row, p. 114.)

Bactericidal action is ordinarily measured quantitatively by viable counts on successive samples of a culture incubated with the drug. These are made by plating serial dilutions on drug-free medium, followed by incubation and then colony counts. Bactericidal action depends on irreversible damage to an essential cell component that cannot be replaced: irreparable damage to DNA; a hole in the membrane too large to seal; or damage to all the ribosomes in a cell, which is irreversible because ribosomal action is required to make more ribosomes.

Bacteriostatic action merely inhibits growth, giving the host's own defenses an opportunity to eliminate the infecting organisms. Many bacteriostatic agents are just as effective as bactericidal ones, except in immunosuppressed hosts or in lesions where the infecting organisms are sequestered from phagocytic cells (as in bacterial endocarditis).

Metabolite Analogs (Antimetabolites)

SULFONAMIDES

The first antibacterial agent used clinically, **sulfanilamide,** and its more potent derivatives (called the sulfonamides) are analogs of an essential metabolite, p-aminobenzoate (PAB). Another analog of PAB, substituted in another part of the molecule, is p-aminosalicylate (PAS), used in the treatment of tuberculosis. By competing with PAB for an enzyme which catalyzes one of the steps in folate biosynthesis, the analogs cause a nutritional deficiency.

The inhibition of bacterial growth by the analogs can be reversed competitively by exogenous PAB; that is, with a doubled concentration of the analog, growth can be restored by doubling the concentration of the substrate. Growth inhibition can be noncompetitively antagonized by a mixture of compounds (methionine, purines, and thymine) whose biosynthesis depends on carbon transfer mediated by the dihydrofolate-tetrahydrofolate cycle. Folic acid (an exogenous precursor of DHF-THF in mammals) is present in body fluids, but most bacteria cannot utilize it; the selective antimicrobial action of sulfonamides depends on this fortunate fact and on the absence of purines and thymine from body fluids. Indeed, the release of these compounds from dead cells makes sulfonamides ineffective in purulent lesions or burns.

OTHER SYNTHETIC ANTIMETABOLITES

The success of sulfonamides led to synthesis of thousands of analogs of known metabolites (amino acids, bases, vitamins). Most of these analogs unfortunately proved just as toxic to animal cells as to bacteria. (Some base analogs, however, were discovered to be useful in cancer chemotherapy.) Antimetabolites of small molecular weights are not limited to chemically synthesized compounds. The structurally simple antibiotic, D-cycloserine, produced by a streptomycete, was found to mimic D-alanine and to block two successive steps in peptidoglycan synthesis. The block can be competitively relieved by D-alanine. Because of its toxicity, cycloserine has little chemotherapeutic value.

Another simple synthetic compound, **isoniazid** (isonicotinic acid hydrazide), is a very valuable agent against *Mycobacterium tuberculosis*. It blocks synthesis of a unique constituent of the mycobacterial wall, the very long chain (C_{80-84}) mycolic acids. Unlike streptomycin, it has the advantages of being effective against intracellular tubercle bacilli, and not selecting for highly resistant mutants. Isoniazid is structurally an analog of nicotinic acid, and also of pyridoxal; the mechanism of action of the drug is not known. Other synthetic compounds, **nitrofurans** and **nalidixic acid,** are not useful for systemic infections but are sufficiently concentrated by the kidney to be useful in urinary tract infections.

Antibiotics

Various antibiotics affect cell-wall synthesis, membrane function or integrity, or protein, RNA, or DNA synthesis; few act on energy-yielding or on central biosynthetic pathways. Those acting on cell-wall synthesis and on protein synthesis are particularly effective in chemotherapy because of the absence of similar targets in the host. The β-lactams, in particular, are remarkably selective, that is, nontoxic to the host.

Antibiotics are best classified by their mechanism of action, which has little correlation with their antimicrobial spectrum. For example, penicillin G, erythromycin, and novobiocin have greatly different structures and actions but almost identical spectra. The spectra appear to be determined by differences in bacterial cell permeability rather than by relative affinities for the target (hence the frequent difference between the sensitivities of gram-positive and gram-negative organisms).

To determine whether an antibiotic acts primarily on the synthesis of DNA, RNA, or protein, one determines the temporal order of the effect of the drug on the incorporation of radioactive precursors (thymidine, uracil, or an amino acid). Synthesis of the primary target macromolecules stops first, followed soon thereafter by the cessation of incorporations into other macromolecules. Action on cell-wall synthesis may be more subtle, affecting the organization of the target rather than the rate of synthesis.

ANTIBIOTICS ACTING ON CELL-WALL SYNTHESIS

Many antibiotics inhibit various steps in peptidoglycan synthesis (Table 13.1), and the resulting accumulations have revealed several biosynthetic intermediates: studies of antibiotic action and of wall synthesis have thus reinforced each other. These antibiotics ordinarily only kill growing cells; growth without accompanying wall synthesis leads to the formation of protoplasts (cells with only a plasma membrane) which readily swell and lyse. However, in isotonic or hypertonic media, the protoplasts (or spheroplasts, which are rounded protoplasts because of total lack of structural constraint by the wall) remain viable.

Penicillins and cephalosporins (β-lactams) The original penicillin was a mixture of several compounds that share a binucleate structure, 6-aminopenicillanic acid (6-APA), but differ in the acyl side chain (Fig. 13.2). This mixture has been replaced in use by the most active component, **penicillin G** (benzyl penicillin), which is particularly effective against streptococci, neisseriae, and treponemas.

The usefulness of penicillin G is limited by its relatively narrow antimicrobial spectrum, its ready destruction by acid in the stomach or by the penicillinase formed by many bacteria, and allergic responses in the patient. The first two problems were solved by the development of a wide variety of semisynthetic penicillins (Fig. 13.2), formed by condensing an acyl chain with the penicillin precursor, 6-APA, which is accumulated by *Penicillium* under certain conditions. Thus, some semisynthetic penicillins are resistant to acid and

Table 13.1. Antibiotics acting on cell-wall synthesis, on membrane, and on DNA synthesis.

Antibiotics	Site	Nature of inhibition	Step of interference
* Penicillins	Wall	Analog of D-Ala · D-Ala bond	Irreversibly block cross-linking of peptidoglycan
Cycloserine	Wall	Analog of D-alanine	Blocks 2 successive reactions of dipeptide synthetase
Phosphonomycin	Wall	Analog of P-enol-pyruvate	Irreversibly blocks formation of enol-ether with UDP-GlcNAc
* Vancomycin	Wall	Blocks lipid carrier, disorganizes the membrane	Blocks subunit transfer from the carrier lipid to the growing peptidoglycan
* Bacitracin	Wall	Blocks lipid carrier, disorganizes the membrane	Blocks dephosphorylation of undecaprenol-PP
* Amphotericin	Membrane	Perturbs structure	Interacts with sterols
* Polymyxins	Membrane	Resemble cationic detergents; bactericidal without growth; probably involves phosphatidyl ethanolamine	Damages membrane, causing leakage
Ionophores	Membrane	Nonpolar rings insert in membrane, the interior of the ring fits specific ion	Increase permeability to specific ion
Gramicidin S Valinomycin	—	—	Highly specific for K+
* Nalidixic acid	DNA replication		Inhibits DNA gyrase activity
* Novobiocin	DNA replication		Inhibits DNA gyrase activity

* Used clinically.

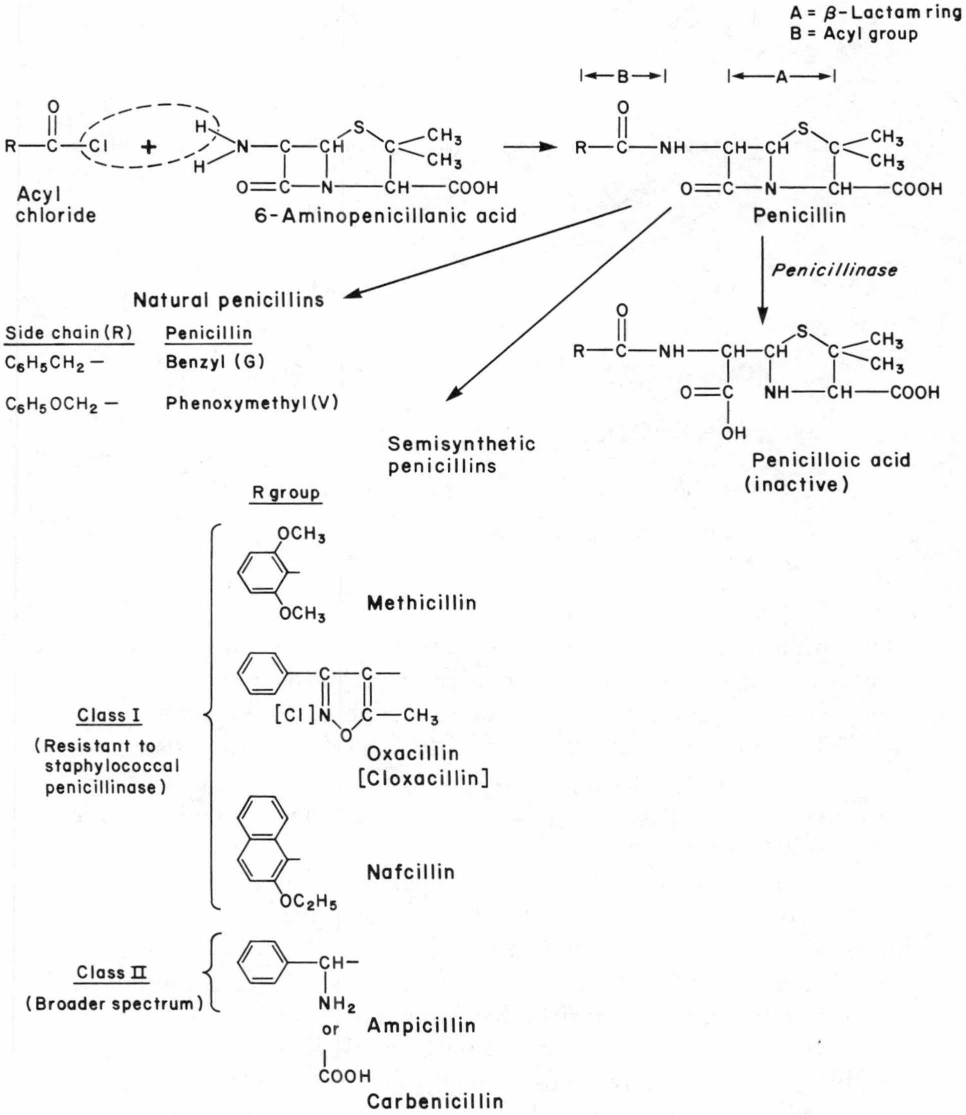

Fig. 13.2 Penicillins: structure and properties.

to penicillinase (for example, **oxacillin**), and others exhibit a broader antimicrobial spectrum **(ampicillin).**

Cephalosporin and its more potent derivatives (for instance, cephalothin) have the same β-lactam ring as penicillin but have an additional carbon in the other ring. Since this latter ring is the main

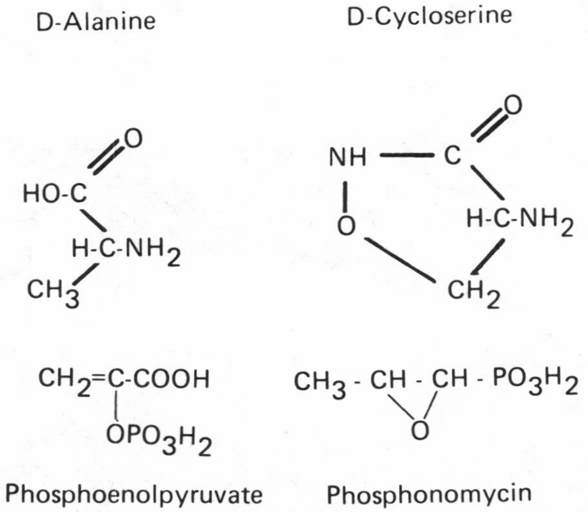

Fig. 13.3 Structural analogy between D-cycloserine and D-alanine, and between phosphonomycin and phosphoenolpyruvate.

determinant in allergy to these compounds, patients allergic to penicillin are usually not sensitive to cephalosporin, and vice versa. Because some penicillinases also act on cephalosporins (and hence are called β-lactamases), semisynthetic cephalosporins resistant to these enzymes and having a wide spectrum as well are often used. Improved semisynthetic products have now been developed for many other antibiotics.

Penicillins and cephalosporins all block the same set of cross-linking enzymes in peptidoglycan synthesis. Some compounds that act on other steps in peptidoglycan synthesis are listed in Table 13.1. Phosphonomycin (fosfomycin) and cycloserine (Fig. 13.3) are not so widely used clinically owing to their toxicity, but they are of interest because of their well-defined action as analogs. Phosphonomycin inhibits the conversion of phosphoenolpyruvate to the lactyl group of the muramic acid in peptidoglycan.

ANTIBIOTICS ACTING ON CELL MEMBRANES

There are also antibiotics that act on cell membranes, some causing gross damage and others increasing permeability to specific ions (Table 13.1).

Polymyxins are polypeptide antibiotics with a polycationic ring and a hydrophobic tail. They are innocuous to membranes of animal cells but bind to bacterial membranes and make them leaky. Gram-negative bacteria are much more susceptible than gram-positive ones, probably owing to the affinity of the antibiotics for lipopolysaccharides. Polymyxins are among the rare antibiotics that are bactericidal even when the cells are not growing.

Polyene antibiotics similarly impair membrane integrity, but do so by complexing with sterols. They are effective against fungi, and also against those mycoplasmas that have incorporated sterols in their membranes.

Ionophores greatly increase the permeability of membranes to specific ions. They are rings that have a nonpolar periphery and insert themselves in membranes. The interior of the ring closely fits a particular inorganic cation. For example, valinomycin specifically encages K^+. These antibiotics are bactericidal, but they are not selective enough to be useful in therapy.

ANTIBIOTICS ACTING ON DNA SYNTHESIS

Novobiocin and **nalidixic acid** are clinically useful antibiotics that inhibit DNA gyrase, an important enzyme in DNA replication and function. Resistant mutants are altered in this enzyme. There are other antibiotics, such as actinomycin D, that bind to DNA and impair its transcription or replication, or cause damage to its structure. These antibiotics are not selective enough to be useful in antimicrobial therapy, but they are of great pharmacological interest because of their use in cancer chemotherapy.

ANTIBIOTICS ACTING ON RNA SYNTHESIS

Antibiotics affecting transcription interfere indirectly with protein synthesis. Their sites of action and mutations leading to resistance are summarized in the upper part of Table 13.2. **Rifamycins** (and the more effective semisynthetic rifampin) block initiation of RNA synthesis by binding to bacterial RNA polymerase. They are not effective against animal polymerases and hence are useful in chemotherapy. RNA chain elongation can be blocked by either streptolydigin, which binds to RNA polymerase, or actinomycin D, which binds to helical DNA at GC pairs.

Table 13.2. Antibiotics inhibiting protein synthesis.

| Antibiotic | Cell type | Site of action | | Altered in resistant mutants | Specific effects |
		Ribosomal subunit	Step		
* Rifamycins	Pro	Transcription		β-subunit	Block initiation of RNA
Streptolydigin	Pro	Transcription		Polymerase	Inhibits extension of RNA
Actinomycin D	Eu, Pro	Transcription		—	Binds to DNA, inhibits extension
Puromycin	Eu, Pro	L	R,P	—	Releases peptidyl-puromycin
* Tetracyclines	Pro	S	R	—	Block binding in A site
* Streptomycin	Pro	S	R	Protein S12	Irreversible; blocks recognition, causing ribosome release; distorts recognition, causing misreading
* Gentamicin	Pro	S/L	R,T	L6	
Neamine	Pro	S	R	S5 & S12 (high level) S17 (low level)	
Kasugamycin	Pro	S	I	S2 or 16S RNA	Reversible; blocks initiation complex formation
* Spectinomycin	Pro	S	T	S5	Probably blocks 1st translocation

* Chloramphenicol, * lincomycin	Pro	L	P	—	Blocks binding to P site, inhibiting puromycin reaction with pp-tRNA
Sparsomycin	Eu, Pro	L	P	—	Blocks peptidyl transfer (puromycin reaction)
Pulvomycin	Pro	S	R		Blocks formation of AA-tRNA · EFTu · GTP
* Erythromycin	Pro	L	P,T	23S RNA, L4 or L26	Blocks P site, inhibiting both peptidyl transfer and translocation on initiating ribosomes
Siomycin, thiopeptin	Pro	L	T,R	— L5	Irreversible; blocks binding of EFG + GTP, and of EFT · AA-tRNA · GTP
Kirromycin	Pro	S	R	EFTu	Blocks release of EFTu and GDP
* Fusidic acid	Eu, Pro	L	T	EFG	Blocks release of EFG and GDP; secondly inhibits A site binding
Micrococcin	Pro	L	T		Blocks translocation
Borrelidin	Pro, Yeast		R		Blocks threonine-tRNA ligase

Eu, Pro = eukaryotic, prokaryotic cells; L, S = large, small subunit; R = recognition; P = peptidyl transfer; T = translocation.
* Used clinically.

ANTIBIOTICS ACTING ON PROTEIN SYNTHESIS

Protein synthesis is the most frequent target of antibiotic action, not surprising in light of the complexity of the ribosome. Most of the useful antibiotics that interfere with protein synthesis, except for the aminoglycosides, are bacteriostatic.

Defining the mechanism of action of an inhibitor of protein synthesis (Table 13.2) entails determining whether the agent binds to the large ribosomal subunit, the small ribosomal subunit, or factors in protein synthesis; whether it blocks ribosomes engaged in chain elongation or only free ribosomes engaged in initiation; and, for an agent that interferes with chain elongation, whether it interferes with recognition, peptidyl transfer, or translocation (see Chapter 5).

Action on the ribosome is more complex than action on one of a sequence of enzymes. Although the ribosome is also a single particle, binding of a ligand to one site very likely affects the conformation or the conformational flexibility of distant sites. Moreover, free ribosomes can bind some antibiotics that polysomes (a chain of ribosomes linked by mRNA during translation) do not bind, because their conformational range is restricted by their ligand. Structures of several antiribosomal antibiotics are shown in Figure 13.4. Some act only on prokaryotes (the basis of therapy), others on eukaryotes, and still others on both systems.

Antibiotics that act only on initiating ribosomes   Erythromycin is a member of the group of macrolide antibiotics which have similar actions; they contain a large aliphatic ring closed by an ester linkage and substituted with sugars. Erythromycin binds 1:1 to free ribosomes and to the 50S subunit but not at all to complexed ribosomes or polysomes. The drug allows initiation to occur, but the resulting initiation complex is blocked. Erythromycin is widely used in therapy; because its binding is reversible, it is bacteriostatic. **Spectinomycin** resembles aminoglycosides in having an aminocyclitol ring but differs in not having amino sugars; it also differs in being bacteriostatic. Like erythomycin, spectinomycin blocks the ribosome as an unstable initiation complex. This drug is important in treating penicillin-resistant gonococci.

Antibiotics that block chain elongation   Puromycin is a structural analog of the aminoacyl end of tRNA (Figure 13.5) and replaces

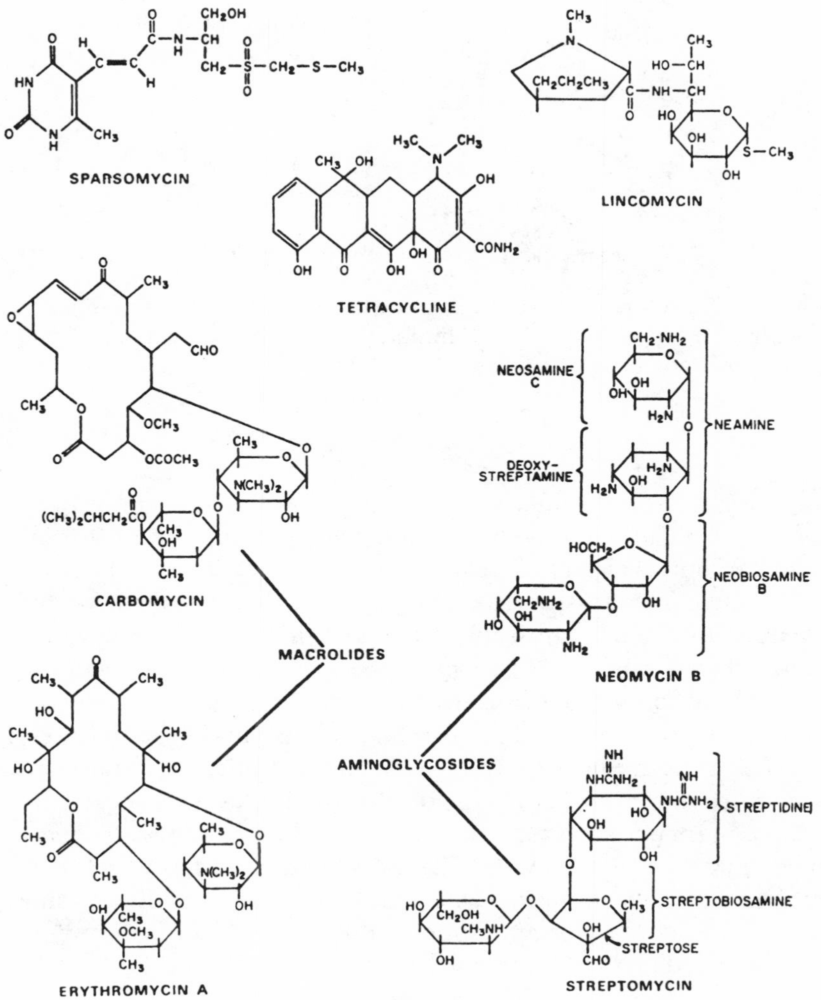

Fig. 13.4 Structures of several antibiotics that bind to ribosomes. (From S. Pestka. 1971. Inhibitors of ribosome functions. *Annual Review of Microbiology* 25:509. © 1971 by Annual Reviews, Inc.)

aminoacyl-tRNA as an acceptor of the nascent peptides from the peptidyl transfer site. Since the peptidyl-puromycin, having much weaker binding than peptidyl-tRNA, dissociates from the ribosome, puromycin inhibits protein synthesis by causing premature release of peptide chains. The ribosomes then fall off the mRNA, and so the

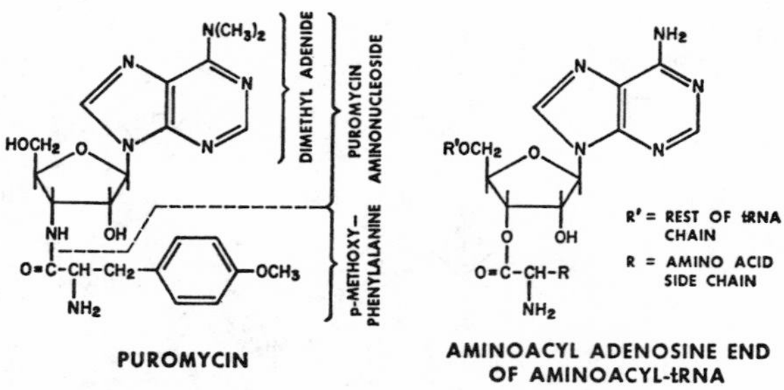

Fig. 13.5 Structures of puromycin and the aminoacyl-adenosine end of AA-tRNA. (From S. Pestka. 1971. Inhibitors of ribosome functions. *Annual Review of Microbiology* 25:516. © 1971 by Annual Reviews, Inc.)

cells lose polysomes and accumulate free ribosomes. Puromycin has been very useful in research but not in chemotherapy because it acts on eukaryotic as well as on prokaryotic ribosomes. **Tetracycline,** one of the original broad-spectrum antibiotics, with a range more extensive than that of penicillin and streptomycin, blocks the recognition step in chain elongation. In the presence of the drug aminoacyl-tRNA cannot bind to its site. **Chloramphenicol,** also an early broad-spectrum antibiotic, binds to the large 50S subunit and blocks peptidyl transfer, thus fixing peptidyl-tRNA in its P site, and stabilizing polysomes in the cell. **Lincomycin** acts much like chloramphenicol in blocking peptidyl transfer in vitro, but its action in the cell appears to be more complex. Its very different structure suggests that peptidyl transfer can be inhibited by agents bound to quite different sites; yet it (and many other antibiotics) inhibits the binding of radioactively labeled chloramphenicol.

Antibiotics with different effects on initiation and on chain elongation After the accidental discovery of penicillin, the first useful product of a deliberate empirical search for antibiotics was made by S. A. Waksman in 1944. This antibiotic, **streptomycin,** a member of the aminoglycoside group, extended chemotherapy to many gram-negative organisms and to the tubercle bacillus. Since then many additional clinically effective aminoglycosides have been found. They include **kanamycin, neomycin, gentamicin,** and **tobramycin,** which have structures and antimicrobial spectra similar to those of streptomycin. All of these drugs have the same bactericidal mode of action on ribosomes.

Having a variety of aminoglycosides available is valuable because they differ in their ability to be inactivated by different plasmid-coded enzymes (see below) and in their toxicity. In addition, except for streptomycin, they bind to multiple sites on the ribosome and hence do not select for one-step mutations to high-level resistance. All are frequently toxic on prolonged administration, and they must be given by injections except for topical treatment. The multiple cationic charges on aminoglycosides appear to be involved in their cellular uptake, for they are antagonized by lowering the pH and by salts—features that impair their usefulness in urinary tract infections. They are also less effective in an anaerobic environment.

Streptomycin and related aminoglycosides are unusual in having two mutually exclusive effects on the ribosome, depending on its state. (1) With *free ribosomes* they cause accumulation of blocked, unstable initiation complexes; but unlike complexes blocked by erythromycin or spectinomycin, the ribosome is irreversibly prevented from subsequent chain elongation in a medium free of the antibiotic. (2) On encountering ribosomes that are already past initiation and are engaged in chain elongation, aminoglycosides have a reversible and less drastic effect. They permit continued (but slowed) translation, but their distortion of the ribosome causes misreading (that is, they relax the discrimination against noncognate aminoacyl-tRNAs). This misreading was first detected (at a sublethal concentration of the drug) as phenotypic suppression of a mutational defect in intact cells. Thus in some auxotrophic mutants, streptomycin restores growth (though it is slow) on minimal medium because an error in translation can occasionally correct an error in the gene, restoring formation of some effective enzyme molecules (but also causing formation of some faulty proteins encoded by functional genes).

Misreading does not appear to be relevant to the chemotherapeutic action of aminoglycosides, but it highlighted the importance of ribosome conformation and revealed the possibility of altering fidelity of translation (which was subsequently also demonstrated by mutations altering the ribosome). In addition, the concept of distortion rather than simple blockage of the ribosome by streptomycin explained a puzzling (and clinically significant) phenomenon: mutations can give rise not only to high-level resistance but also to dependence on streptomycin for growth (which has been traced to an altered protein S12 of the 30S subunit). Distortion by the antibiotic evidently compensates for the mutational deformation in the ribosome and restores normal codon-anticodon alignment.

Antagonism and Synergism

The bactericidal action of aminoglycosides is slowed by poor nutrition. However, what is required for this action is not growth or protein synthesis per se but is evidently ribosomal recycling, since killing is prevented by chloramphenicol (which fixes the polysomes)

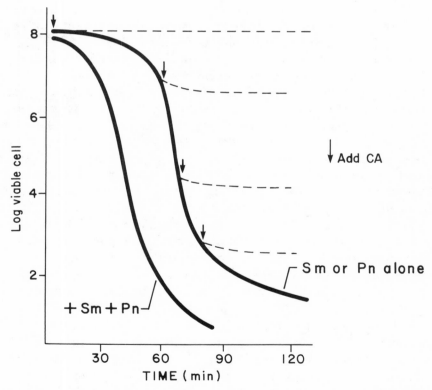

Fig. 13.6 Antagonistic protection by chloramphenicol (CA) against killing by streptomycin (Sm) or by penicillin (Pn), and synergistic killings by streptomycin and penicillin. Solid lines represent the decline in viability and dashed lines represent the arrest of killing.

but not by puromycin (which allows recycling but prevents protein synthesis). Antagonism by bacteriostasis is a strong argument against mixed treatment with an aminoglycoside and a growth-inhibiting agent. The same holds for β-lactams (Fig. 13.6), since their lytic action requires substantial growth and cell-wall synthesis.

In contrast, β-lactams and aminoglycosides are synergistic in killing bacteria. The β-lactams, in exposures that damage the cell envelope but are not yet lethal, facilitate uptake of the aminoglycosides. As shown in Figure 13.6, exposure first to penicillin and then to streptomycin is synergistic. (The reverse sequence is not.) This strategy has proved especially valuable in eliminating the very sluggishly growing organisms in subacute bacterial endocarditis.

Drug Resistance

MECHANISMS

A strain may acquire drug resistance (so that an increased concentration of the drug is necessary for inhibition or killing) either by mutation or by gene transfer (see Chapter 8). The genetic alteration may cause resistance by various physiological mechanisms.

Decreased permeability   Certain strains achieve resistance to amino acid analogs, aminoglycosides, or tetracycline by decreasing uptake of these agents. Low permeability is probably the main basis of the "natural resistance" that limits the antimicrobial spectrum of various agents.

Enzymatic inactivation   Antibacterial agents can also be enzymatically inactivated by the infecting organisms. For example, such inactivations may result from: (1) hydrolysis (penicillin), (2) phosphorylation, acetylation, or adenylylation (aminoglycosides), and (3) acetylation (chloramphenicol, sulfonamides). Resistance due to extracellular inactivating enzymes, such as the excreted penicillinase of some organisms in case 1, depends on population density of the cells; hence it does not have a fixed numerical value. In contrast, resistance due to intracellular enzymes (which require intracellular metabolites such as ATP or acetyl CoA), as in cases 2 and 3 or cellular β-lactamase, is relatively independent of population density. Inhibition of growth would depend on the rate of entry of the drug versus its rate of inactivation.

Target site alteration   Resistance may also result from an alteration that renders the target site less sensitive to the drug. Examples include a decreased affinity of ribosomes for streptomycin (traced to a change in protein S12) and a decreased affinity of the target enzyme for sulfonamides. Single mutations to streptomycin resistance can alter the level a thousand-fold (making the ribosome resemble mammalian ribosomes in this respect).

Inducible resistance   In some cases resistance can be phenotypically increased by exposure to the drug. Subinhibitory concentrations of erythromycin apparently induce an enzyme that methylates the 50S rRNA, thereby increasing the resistance to the drug. Likewise, the export system for tetracycline encoded by the transposon Tn10 is also inducible by the drug.

Other mechanisms   Resistance may be a consequence of decreased activation of a drug (for example, purine or pyrimidine analogs that must be converted to nucleotides before they can interfere with essential reactions), increased production of an antagonistic metabolite (for example, PAB versus sulfonamides), and increased production of the target enzyme (for example, dihydrofolate reductase in mammalian cells resistant to methotrexate).

PHENOTYPIC RESISTANCE

Typically, the killing of growing microorganisms in cultures by a bactericidal drug is initially exponential, but it often levels off, leaving a few **persisters,** cells that are not growing but can eventually resume growth after removal of the drug. In contrast to the clones derived from a rare resistant mutant (which can be said to have acquired genotypic resistance), the clones derived from persisters are as sensitive to the drug as the original organism.

The basis for the survival of the persisters is difficult to study. The current view is that in a large population, there are always some cells which are in a state of dormancy because of reparable defects. As a rule, "resting" cells are less vulnerable to antibiotics. Perhaps the effectiveness of some antibiotics depends on their active transport into the cells.

Persisters contribute to the usual failure of chemotherapy to be curative when it is applied only briefly. In the body, bactericidal action is often slowed or prevented by additional mechanisms such as the sequestering of bacteria in locations that are not supportive of growth (because of inadequate nutrition or supply of O_2, or accumulation of growth inhibitors).

Eradication of persisters, like the eradication of cells that are inhibited by bacteriostatic agents, requires immune and phagocytic responses. Fortunately, these bactericidal responses, directed to the bacterial surface, do not depend on metabolic activities of the bacteria and hence can proceed under circumstances that impair antibiotic bactericidal action.

PREVENTION OF RESISTANCE

Streptomycin and some other agents can select for single-step mutations to high-level resistance, but with most agents resistance by mutation develops in small increments. In those cases, treatment with high doses would hinder the emergence of drug resistance. Combination therapy with two agents that attack different targets is another good strategy, since a mutant resistant to one drug will still be susceptible to the other.

Conclusion

Studies on antibiotics that attack bacteria with minimal harmful effects on their hosts revealed critical differences that culminated in the divergent evolution of prokaryotic and eukaryotic cells. The development of peptidoglycan by bacteria for protection against osmotic stress renders them especially vulnerable to chemicals that interfere with the synthesis of this polymer. The absence of sterols in the phospholipid membranes of the vast majority of bacteria and the presence of specialized proteins on these membranes also allow differential drug action. Finally, the specialization of the machineries for replication, transcription, and translation provides many more opportunities for selective targeting against pathogenic bacteria.

Selectivity is probably sometimes blunted, however, by the presence in eukaryotic cells of organelles of prokaryotic origin. For example, mitochondria are believed to be the evolutionary descendants of prokaryotic endosymbionts specializing in oxidative phosphorylation. These organelles contain their own small circular genome and have 70S instead of 80S ribosomes. It is therefore not surprising that protein synthesis in mitochondria is inhibited by chloramphenicol. This might account for the drug's adverse or even fatal effects on the hematopoietic system, but it is not at all clear why this system should be particularly susceptible.

In comparison with the processes of informational transfer and macromolecular assembly, the degradative and biosynthetic pathways for small molecules are more universal and the proteins involved in the catalyses are less complex. For these reasons relatively few agents can be expected to act at the level of intermediary metabolism. Thus the fortunate combination of circumstances that made sulfonamides successful might well be a unique phenomenon. Indeed, the thousands of analogs of amino acids, bases, and vitamins synthesized for potential antibacterial chemotherapy have proved to be useless because of damaging effects to the host.

Current therapeutic approaches focus on attacking the molecular mechanisms that confer on the bacteria their specifically pathogenic properties, such as the production of cholera toxin or the elaboration of special appendages that allow the bacteria to attach to host cell surfaces. Successful treatments by these approaches have the added advantage of not disturbing the normal flora, many of which ordinarily live in peace or in a symbiotic relationship with the host.

Questions

13.1. A strain of *Mycobacterium tuberculosis* is isolated from a patient. In the antibiotic sensitivity test it is found to be resistant to streptomycin. (a) How is the test conducted? (b) What might be the basis for the resistance? How might it be determined? (c) How should this patient be treated?

13.2. A strain of *Staphylococcus aureus* is known to possess a plasmid (not readily transmissible) that confers resistance to penicillin and spectinomycin. A strain of *E. coli* is resistant to streptomycin and tetracycline but contains no plasmid. (a) In a growth-supporting medium containing 10^7 cells/ml of each organism, how will the viable number of each kind change after one hour in the presence of the following antibiotic combinations? Explain. (Assume the generation time of each organism is one hour, and ignore the rare mutants that may exist.)

(1)   Streptomycin and tetracycline.
(2)   Penicillin and tetracycline.
(3)   Spectinomycin and tetracycline.
(4)   Spectinomycin and penicillin.

(b) How would one determine the numbers of each organism in the mixture?

13.3. At what sites of action can antibiotics be expected to have lethal effects?

13.4. A clinical isolate of *E. coli* was obtained from a patient with diarrhea and found to be resistant to the antibiotic ampicillin. Which one of the following statements is definitely *not* the explanation for this resistance?

(1)   The strain produced a chromosomally encoded β-lactamase which inactivates the ampicillin.
(2)   Ampicillin is no longer permeable across the cytoplasmic membrane.
(3)   The enzyme which is the target of ampicillin has been altered such that it no longer binds ampicillin.
(4)   The strain produces a β-lactamase encoded by a resistance factor.

13.5. (a) Explain the synergistic effect in the treatment of enterococci with penicillin and the aminoglycoside streptomycin. (b) Explain how high and low level resistance to streptomycin can occur. (c) Explain streptomycin dependence.

13.6. A dose of 10^5 *Salmonella* cells given orally causes diarrhea in a healthy volunteer. But pretreatment of the volunteer with streptomycin decreases to 10^3 the number required to cause diarrhea. Explain.

Further Reading

Aoki, H., and M. Okuhara. 1980. Natural β-lactam antibiotics. *Annual Review of Microbiology* 34:159–181.

Chopra, I., and P. Ball. 1982. Transport of antibiotics into bacteria. *Advances in Microbial Physiology* 23:184–240.

Davis, B. D. 1982. Bactericidal synergism between β-lactams and aminoglycosides: mechanism and possible therapeutic implications. *Reviews of Infectious Diseases* 4:237–245.

Pestka, S. 1977. Inhibitors of protein synthesis. In *Molecular Mechanisms of Protein Biosynthesis*, ed. H. Weissbach and S. Pestka. New York: Academic, pp. 467–553.

Stuart-Harris, C. H., and D. M. Harris. 1982. *The Control of Antibiotic-Resistant Bacteria*. New York: Academic.

Tomasz, A. 1979. The mechanism of the irreversible antimicrobial effects of penicillins: how the β-lactam antibiotics kill and lyse bacteria. *Annual Review of Microbiology* 33:113–137.

Answers to Questions

Index

Answers to Questions

2.1. (a) "Protective conditions" means a medium of high osmotic strength, so protoplasts will not lyse. (b) In local lesions there might be such environments. (c) The protoplasts no longer produce targets for the drug. (d) Some protoplasts are so denuded that they no longer have peptidoglycan primer to accept the disaccharide units, so that the cell wall might not be able to regenerate even in the absence of the antibiotic. (There are other possibilities. For instance, growth without peptidoglycan apparently favors mutants with early blocks in the pathway.)

2.2. (a) Enzymes such as alkaline phosphatase are nonspecific in their action and would destroy essential phosphorylated metabolic intermediates in the cell; hence their mechanism of synthesis places them outside the cytoplasmic membrane. (b) Gram-positives have no periplasmic space. Apparently the layers of peptidoglycan cannot effectively retain the exoenzymes. (c) Enzymes in the periplasm might be more useful to the organism than those that are excreted and diluted into the environment, provided that the substrates are not too large to enter through the pores of the outer membrane. In contrast, excreted enzymes would also benefit the nonproducers in the neighborhood.

2.3. (a) Mutants lacking O-antigen may have a more hydrophobic outer membrane, making penicillin G (hydrophobic) penetrate with greater ease. EDTA alters the outer membrane of wild-type cells by chelating Mg^{++} and releasing some LPS. Thus the drug treatment has an effect similar to the mutation. (b) Actinomycin, being hydrophobic, will also enter the modified cells faster. (c) Lysozyme, too, despite its large size, will have more ready access to the peptidoglycan. (d) It would make little difference to small molecules like cyanide.

2.4. The answers may be obtained by proceeding from data given in the table line by line according to the principle of Ockham's razor:

(a) X and Z share an antigen I, but no antigen is shared by X and Y. Hence: X (I^+); Y (I^-); Z (I^+).

(b) Antiserum to X agglutinates Z because of *specific* antibodies against X.

(c) Confirms that X shares no antigen with Y, since the agglutinating power is not diminished.

(d) X has an antigen II not shared by Z. Hence: X (I^+,II^+); Y (I^-,II^-); Z (I^+,II^-).

(e) Y shares an antigen III with Z. Hence: X (I^+,II^+,III^-); Y (I^-,II^-,III^+); Z (I^+,II^-,III^+).

(f) Confirms Y shares no antigen with X.

(g) Antiserum to Y agglutinates Z because of *specific* antibodies against Y.

(h) Y has an antigen IV not present on Z. Hence: X (I^+,II^+,III^-,IV^-); Y (I^-,II^-,III^+,IV^+); Z (I^+,II^-,III^+,IV^-).

(i) Z has a common antigen with X and a common antigen with Y. This can be explained by the above assignment.

(j) Z has a common antigen with X, which would be I^+.

(k) Z has a common antigen with Y, which would be III^+.

(l) Antiserum to Z agglutinates X and Y because of *specific* antibodies against Z. Conclusion: X (I^+,II^+); Y (III^+,IV^+); Z (I^+,III^+).

2.5. Synthesis of peptidoglycan probably involves rearrangement of various chemical bonds and the shuffling of DAP to accommodate cell growth.

3.1. Specialized animal cells, such as intestinal epithelial and kidney tubule cells, help to assure the rest of the cells a more or less uniform environment containing adequate concentration of amino acids, sugars, and so forth, so that in most cases active transport is not necessary.

3.2. The growth would be slow when the concentrations of lactose are low, but otherwise normal.

3.3. One might expect growth to be poor on almost all hydrophilic small molecules unless they had a different way of passing through the outer membrane (maltose, for example, induces its own outer membrane pore).

3.4. Since CCCP collapses the proton gradient, we infer that TMG uptake in *S. lactis* is driven by the proton motive force. In other streptococci, TMG is trapped by the phosphotransferase system where energy is provided by substrate-level phosphorylation (that is, with phospho-enolpyruvate).

3.5. Many wild-type cells in the center will not "sense" any concentration gradient because the compound is completely depleted in the central area. Therefore, there will be no stimulus for them to migrate. Also, any nonmotile or nonchemotactic mutants will fail to migrate.

3.6. Anaerobically, the $2 \times 2H$ derived from glucose must be disposed of by reduction of acetyl-P to ethanol via acetyl-CoA. This requires the sacrifice of a high-energy bond; therefore ATP has to come from the pathway:

$$\text{Glyceraldehyde-3-P} \underset{2H}{\overset{2H}{\rightleftarrows}} \text{pyruvate} \longrightarrow \text{lactate.}$$

But aerobically, the hydrogens react with O_2 to give H_2O_2, so acetyl-P can be used instead to generate ATP by substrate level phosphorylation. Hence, 2 molecules of ATP are made per molecule of glucose, instead of 1. (Admittedly, this does not explain why lactate is still made.)

3.7. Yes. In such an organism, there occur no O_2-using reactions of the type: $XH_2 + O_2 \rightarrow X + H_2O_2$. So no superoxide is formed.

4.1. (a) The volume of this hemisphere is $\frac{1}{2} \times \frac{4}{3}\pi \times (100\,\mu)^3$; and the volume of the cell is $\pi \times (0.25\,\mu)^2 \times 1\,\mu$. This gives about 10^7 cells in the hemispheric colony. (b) It takes about 23 divisions to give 10^7 cells and this takes place within 10 hours. Therefore the doubling time is about $\frac{1}{2}$ hour. The doubling time can also be calculated from:

$$\ln N_2/N_1 = k(t_2 - t_1)$$

where k in this equation is the first-order growth constant.

4.2. By serial dilution until the average titer is 0.1 cells/ml. By inoculating 20 tubes of culture medium with 1 ml of diluted bacteria, on the average 2 tubes will contain a single cell. With lesser dilutions there is a high probability of inoculating two cells in the same tube.

4.3. About 10^{10}/ml.

4.4. Three days would allow 72 divisions, $2^{72} \times 10^4 = 4.7 \times 10^{25}$. Note that:

$$\log(2^{72}) = 72 \times \log 2 = 72 \times 0.3010 = 21.67$$
$$10^{21.67} = 10^{21} \times 10^{0.67} = 4.7 \times 10^{21}.$$

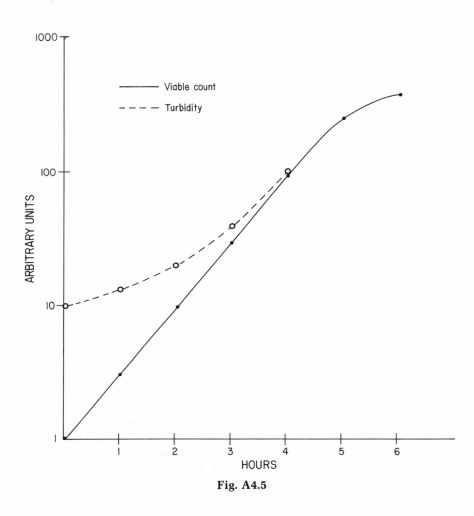

Fig. A4.5

4.5. See Figure A4.5.

4.6. See Figure A4.6.

4.7. Reasonable explanations are a, c, d.

4.8. Butts are yellow because fermentation of glucose gives acid products which are not reoxidized under the semi-anaerobic condition. The slant on which *Salmonella* has grown is red because lactose is not used and the acid produced from glucose is oxidized. The slant on which *E. coli* has grown is yellow because this organism is able to use the abundant lactose (10 times the concentration of glucose); too much acid

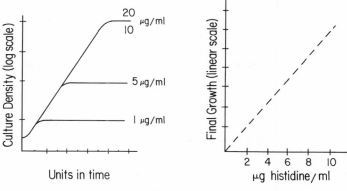

Fig. A4.6

is made to be reoxidized in the time of incubation. The precipitation test indicates that *Salmonella* has a thiosulfate cleavage enzyme and that *E. coli* does not.

5.1. (a) Mutagenize a population of cells; allow them to grow out in a simple defined medium supplemented with threonine and methionine (or homoserine); enrich the mutant population by killing the wild-type cells with penicillin (or a drug of similar action) in a medium without the growth factors. Step 3 is affected by the mutation. The mutant needs to be fed only threonine and methionine (or homoserine). Isoleucine is not needed since it can be derived from threonine.

(b) Step 9 (because an earlier block would impose diaminopimelate requirement).

(c) Decreased affinity of amino-acyl synthetase for isoleucine, for example.

5.2. (a) On the basis of these experimental results, the pathway has *at least* 3 genes. (b) Mutant A is blocked in step 1, C and D in step 2, and B in step 3 as shown below:

$$X \xrightarrow[1]{A} Y \xrightarrow[2]{C,D} Z \xrightarrow[3]{B} \text{histidine.}$$

Reasoning: mutants C and D accumulate intermediate Y, which is excreted and fed to mutant A, and so on. (c) The wild-type strain will not feed any of the mutants because histidine controls its biosynthesis by feedback so there is no excess to cause excretion.

5.3. The proteins for lactose utilization have to be highly induced before steady-state growth can resume. The rate of induction is dependent on the basal permease and β-galactosidase levels present in cells grown on glucose, since these basal levels determine the rate at which the true inducer, allolactose, is formed. The basal expression of the lactose genes might be strain-dependent. Proteins for glucose utilization are either constitutive or present in high basal levels in all strains of E. coli, so rapid utilization of the sugar can occur irrespective of the previous growth condition.

5.4. (a) The PG-utilizing cells are most likely to be mutants with constitutive expression of the lac operon, because enzyme induction is no longer necessary after transfer from a glucose to a lactose medium. Apparently PG is a substrate for both β-galactose permease and β-galactosidase, but the compound cannot be served as an inducer. (b) The TONPG-tolerant cells very likely lost β-galactoside permease (or both the permease and β-galactosidase). If so, they should be lactose-negative. The experimental observations also suggest that TONPG is a substrate for β-galactoside permease but not an inducer and that accumulation of this compound (or its product) by cells constitutive in the lac operon is toxic. (Actually, TONPG itself is accumulated; the analog is not hydrolyzable by β-galactosidase because of the sulfur bridge.)

6.1. Mutations that result in changing a codon to its synonymous codon (specifying the same amino acid); mutations that substitute the original amino acid with another one having similar property (for example, valine to leucine); mutations that result in the alteration of an amino acid not occupying a "critical" position in the polypeptide.

6.2. (a) Mutation to UUG (leu) probably has little or no effect on the function of the gene product, but tolerance of such a change is likely to be dependent on the location of the amino acid in the polypeptide chain. So no strong statement can be made. Mutation to UGG (trp) might be more consequential than to UUG (leu) on stereochemical grounds. Mutation to UAG would be most drastic, since this causes premature chain termination in the translational process.

(b) 2-Aminopurine induces A to G (also U to C) and G to A (C to U) transitions. Therefore, the leucine and serine phenotype can be produced. Hydroxylamine induces C to U (G to A) transitions. Therefore, the leucine and serine phenotype can also be produced. No effect is expected of acridine orange since this mutagen promotes frameshift.

6.3. First, there must be redundant genes coding for the tRNA. A mutation to UAG can be suppressed by a mutation in the third position of the anticodon of a redundant tRNA for tyrosine. Second, UAG is not the most frequently employed termination codon. Otherwise the suppressor tRNA would seriously interfere with translational termination.

6.4. Possibly d. This will delete three amino acids and restore the reading frame; these three amino acids must be dispensable.

6.5. Mutant 1 has sustained a deletion mutation. Mutant 2 has sustained a transition mutation. Mutant 3 has sustained a nonsense mutation. Mutant 4 has sustained a frameshift mutation.

6.6. Mutations in a gene specifying an enzyme should sometimes yield proteins with altered properties, such as increased thermal lability (temperature-sensitive mutants) or defective catalytic activity.

6.7. (a) A missense mutation in the $lacZ$ gene, resulting in a protein without catalytic activity. (Acridine would stimulate reversion only if the forward mutation was a frameshift.) (b) A deletion affecting the $lacZ$ gene, confined within this gene or extended into the $lacY$ gene, which might create a nonsense codon on the distal side with polarity effect. (c) A small deletion at the end of the $lacZ$ gene affecting the carboxyl terminus of the protein.

6.8. (a) The lack of cAMP would affect the expression of all operons that are subject to catabolite repression. (b) "Revertants" that grow on lactose, but not on the other sugars like maltose and arabinose, might result from an alteration of the cAMP binding site of the lac operon. Lactose-positive revertants may also result from large genetic rearrangements that juxtapose the lac structural genes to some other promoter that is active without cAMP. Such genetic rearrangements could involve transposition, inversion, deletion, or insertion.

6.9. Nonsense mutations in the ara-3 gene (see Fig. 5.6) might give a similar phenotype by polarity effect. If the nonsense mutation is in the regulatory gene, a partial diploid $araR^-$, $ara3^+/araR^+$, $ara3^-$ should usually become inducible and be able to grow on arabinose. The wild-type $araR^+$ allele should show trans-dominance in permitting the expression of the $ara3^+$ gene in the other genome. The mutation may also be in the gene controlling arabinose transport, which happens to be unlinked to the arabinose operon. If this is true, it could be identified by genetic mapping.

6.10. (a) Any termination codon (UAG, UAA, UGA) causes a protein release factor to hydrolyze the bond between the polypeptide and the ultimate tRNA, thus allowing the ribosome to fall off. The absence of ribosome stalling would give a false signal of his-tRNA abundance and reduce the rate of transcription of the structural genes for histidine biosynthesis by attentuation. (b) The mutation should be cis-dominant.

6.11. If both genes belong to the same operon, nonsense mutations in one of the genes (proximal to the promoter) should disrupt the function of the other gene (distal to the promoter).

6.12. Insertion of IS2 in one orientation may introduce a ρ termination signal, whereas in the second it does not. Alternatively, insertion in one orientation may show no polarity, because a new promoter is provided for the expression of the distal genes, without nonsense codons in the intervening sequence. Thus, an insertion is tolerable only if neither transcription nor translation is functionally disturbed.

6.13. It is quite possible that such a mutation is tolerable, and the only consequence would be a longer transcript. But the absence of ρ activity is lethal, because extensive loss of transcription termination is intolerable, or perhaps, ρ is involved in regulating some essential process.

7.1. (a) In such a merodiploid cell (containing a DNA fragment with an S gene) capsule would be made. However, without recombination into the chromosome, the S gene would be passed to only one progeny cell at each division, since the fragment is unlikely to possess the site that acts as the origin for DNA replication. (b) Eventually, this kind of DNA would be destroyed by nucleases. Therefore the cell will not give rise to a small colony.

7.2. (a) The isolate might be tested for the ability to conjugate with a known F^- recipient and for the ability to transfer chromosomal genes. Further evidence might be obtained by screening for a plasmid by physical techniques. The isolate might also be screened for sensitivity to a bacterial virus using F pilus as receptor. (b) Cells harboring the F plasmid might grow slower because of the extra DNA with its associated metabolic burden. These cells might also be selected against by male-specific bacterial viruses. Under certain conditions, such as low temperature, the replication of the plasmid might not keep up with the replication of the chromosome. There might be other reasons.

7.3. A wild-type phenotype can be restored by recombination as a result of interchange of parts between two different mutant DNA molecules by homologous pairing in such a way that one recombinant DNA molecule emerges without defect. Complementation restores wild-type phenotype by having in the same cytoplasm two different mutant DNA molecules, each with a defect in a different gene, so that the presence of one copy of a functional allele compensates for the inactivity of the corresponding allele. Recombinational analysis can determine the relative positions of two different mutational sites; complementational analysis can determine whether or not two mutations affect the same gene.

7.4. (a) General methods: examine conditional lethal mutants that fail to grow at high temperature under any nutritional condition, or mutants that require the presence of a nonsense suppressor. Special methods: examine mutants that are resistant to RNA polymerase inhibitors such as rifamycin, rifampin (a semisynthetic derivative), or streptolydigin (see Table 13.2). (b) Mutations affecting different subunits can be revealed by complementation analysis.

7.5. (a) The two mutations (markers) that are farthest apart are 3 and 4 (1.5 recombination units). Therefore, 1 and 2 must be flanked by 3 and 4. Moreover, 1 and 2 are very close to each other (0.05 units). The relative positions of 1 and 2 are determined by the following data: The distance between 1 and 4 (1.0 unit) is greater than the distance between 2 and 4 (0.9 units). This is consistent with the observation that the distance between 2 and 3 (0.8 units) is greater than the distance between 1 and 3 (0.7 units). The answer is shown in Figure A7.5. In experimental reality, the data might not be so neat.

(b) There are two complementation groups, that is, two genes.

(c) Trp^-4 does not provide any diffusible gene product that can compensate missing gene products of the other mutants, and yet does not prevent the proper function of a set of wild-type genes. Hence, this

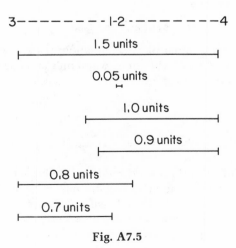

Fig. A7.5

mutation might interfere with the transcription of an mRNA that codes for all the genes required for tryptophan synthesis.

(d) This is probably a mutation not linked to the cluster of genes specifying tryptophan synthesis. Therefore a wild-type copy of the gene is not carried by the F'-trp^+ plasmid. For example, the mutation might be in the gene coding for aminoacyl-tRNA synthetase for tryptophan causing reduced affinity of the enzyme for tryptophan, so that the intracellular concentration of the amino acid required is higher than can be attained by synthesis. (Recall from Chapter 5, the tryptophan operon is under both repressor and attenuation feedback control.) Presence of exogenous tryptophan would elevate the internal concentration of the amino acid, because there is an active transport system.

7.6. X must be within the narrow sector in the middle of gene C since this is the only region that *overlaps* with deletions 2, 3, and 4. Hence, there is no wild-type recombination.

7.7. A true revertant should produce a gene product that is indistinguishable from the wild-type product and at the same level. Moreover, back cross with the wild-type should not give mutant segregants. A suppressor mutant may produce a gene product that is detectably different from a wild-type product and/or at an altered level. Moreover, back cross with the wild-type should give rise to mutant segregants.

7.8. Rough mutants are engulfed by macrophages without need for specific antibody, but smooth cells are not.

7.9. (a) Noninducible, (b) constitutive, (c) inducible, (d) constitutive.

8.1. 2.5×10^{-27} or immeasurably low.

8.2. Inversion of the markers *sul* and *cml* (see Fig. 8.5).

8.3. (a) After a single round of recombination, there are 3 possible outcomes (see Fig. A8.3a). (b) After two rounds of recombination, only 2 possible structures in addition to the parental plasmid can be formed (see Fig. A8.3b).

8.4. See Figures A8.4a,b,c.

8.5. (a) Repeated laboratory passage of strain H apparently diluted out the plasmid that enables the cell to produce the pilus which is probably critical for adhesion to the gut mucosa. Synthesis of these structures under laboratory conditions is a gratuitous metabolic burden. Therefore, it is not surprising that cells without the plasmid would have a growth advantage. Elimination of this extrachromosomal element without special measures being taken to prevent reinfection in the

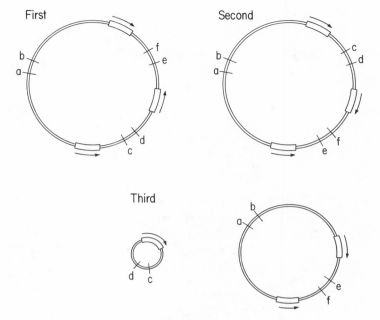

Fig. A8.3a

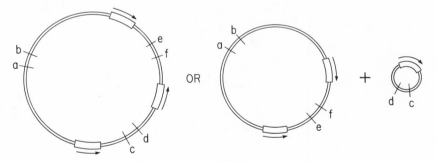

Fig. A8.3b

culture indicates that the plasmid does not code for a conjugative apparatus.

(b) If the plasmid actually codes for the structure of the pilus, rather than a regulatory element that turns on pilus synthesis, one should be able to demonstrate this by first obtaining a mutant bacterial strain that produces an altered pilus. The plasmid can then be isolated from such a mutant and used for transformation of strain H-P. If the structural gene

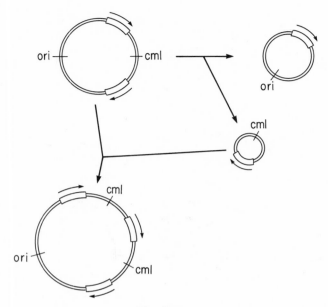

Fig. A8.4a

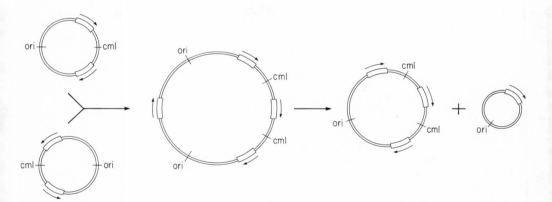

Fig. A8.4b

for the pilus is plasmid-determined, the transformed bacteria should produce the novel pilus. The plasmid might code for other cellular functions in addition, for example, special metabolic abilities. These functions in general have to be identified empirically.

9.1. (a) Adsorption. Grow the phage mutant at 30°C in the presence of

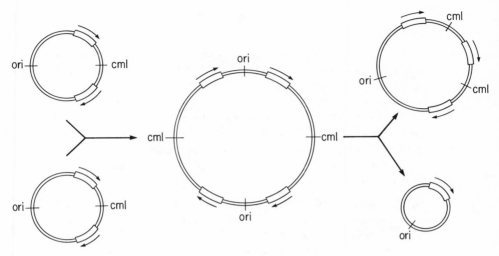

Fig. A8.4c

^{3}H-thymidine to label its DNA and ^{14}C-leucine to label its capsid. The labeled phages are then mixed with unexposed cells at 42°C. After a suitable time for normal adsorption to occur (10 min), the cells are pelleted by centrifugation at a speed that will not sediment unadsorbed phage. The pellet and the supernatant fluid are then examined for radioactivity. Presence of most of the ^{3}H- and ^{14}C-labels in the supernatant fraction would indicate defective adsorption.

(b) In the above experiment, if most of the ^{3}H and ^{14}C-labels associate with the cells, the pellet can be resuspended and subjected to mechanical disturbance (for example, agitated in a Waring blender). The phage capsids sheared off can be separated from the cells by light centrifugation. Retention of the same ^{3}H/^{14}C ratio in the capsids would indicate failure of DNA injection.

(c) Defective DNA replication can be revealed by incubating infected cells under nonpermissive conditions in the presence of labeled thymine and examining for the increase in phage DNA.

(d) Defect in the synthesis or assembly of structural subunits of the phage can usually be revealed by electron microscopy of a lysate (chloroform treatment) of infected cells. One can look for the presence or absence of capsids, tails, and so on.

(e) Release. Cellular lysis can be monitored by the optical density (absorbancy) of an infected culture over time. If lysis does not occur at the usual time, one might artificially lyse the culture with chloroform and test for plaque-forming units on a lawn of host cells at the permissive temperature.

9.2. (a) More than one phage can attach to a bacterial cell, so that each cell that adsorbs more than one phage will spare another cell in the population from infection. When this culture is diluted, the uninfected cells will grow up to give a colony. The fraction of uninfected cells is about 30% (according to Poisson distribution). Since all cells infected with T4 are killed, the surviving fraction is 3×10^9.

(b) λ is temperate phage; in this case many infected cells survive by becoming lysogens. (A fraction of the cells might survive because of abortive infection.)

9.3. (a) The infected culture will regain turbidity after overnight incubation, because the few spontaneous phage-resistant cells among the large original population will have time to grow out. (b) Resistance to two different virulent phages are likely to require two independent mutations. The frequency of such double mutants is extremely low. The population of the original bacterial culture is evidently not large enough to provide such a mutant.

9.4. (1) One of the genes codes for a subunit of a multimeric protein, say α_4. The mutant gene directs the synthesis of an α^*, such that not only is α_4^* inactive, but the presence of any α^* in the multimer (that is $\alpha_1^*\alpha_3$, $\alpha_2^*\alpha_2$, and $\alpha_3^*\alpha_1$) will render the complex inactive. Thus only the small fraction of pure wild-type α_4 retains activity, but this is insufficient to assure phage growth. (This phenomenon is referred to as negative complementation.)

(2) The two affected structural genes belong to the same operon. One phage has a nonsense mutation in the upstream gene near the promoter. The polar mutation therefore abolishes the formation of both gene products, making complementation with the other phage (with a mutation in the downstream gene) impossible.

(3) One mutant might fail to adsorb or inject its DNA into the host cell.

(4) One mutation might be dominant.

10.1. The mixed progeny will include (1) T4 DNA in T4 coat, (2) T4 DNA in T2 coat, (3) T2 DNA in T4 coat, (4) T2 DNA in T2 coat. The "coat" in this context includes the tail and tail fibers which will determine host range. (a) All phage progeny from the above infection will give plaques. (b) Since strain B/2 only has membrane receptors for the T4 coat, only phage particles with the T4 exterior will adsorb and subsequently infect the cell. But, T2 DNA in T4 coat will not give rise to a plaque when plated on a B/2 lawn. (c) T4 DNA in T4 coat will give clear plaques. T2 DNA in T4 coat will give turbid (or invisible) plaques. T4 DNA in T2 coat will give clear plaques. T2 DNA in T2 coat will give turbid (or invisible) plaques.

10.2. These results would suggest (but not prove) that a membrane component of the B-12 transport system serves as a receptor or a DNA injection site for bacteriophage BF23.

10.3. (a) The mutant gene product should differ from the wild-type gene product by one amino acid when made in the suppressor$^+$ host, unless by coincidence the suppressor tRNA causes the incorporation of the original amino acid.

(b) Amber suppression can only restore the synthesis of a functional protein to a subnormal level (in this case 20%), since the suppressor tRNA competes with the termination factor. Proteins with catalytic activity are generally needed in fewer copies than proteins with structural function. This would explain why the suppression of a mutation in a gene specifying an early function (usually a gene coding for an enzyme) is apparently more effective than the suppression of a mutation in a gene specifying a late function (usually a gene coding for a structural component of the phage itself and therefore needed in a stoichiometric quantity).

10.4. Deployment of host polymerase for transcription of phage genome at the expense of host function.

10.5. Ribosome translocation along the mRNA in the 5′ to 3′ direction terminates at the nonsense codon. Nonsense mutations early in the coat protein gene allow ribosomes to fall off prematurely, thus leaving the RNA molecule locally folded in its native conformation. The ribosome-binding site of the replicase gene in this phage is buried and inaccessible to translation unless the RNA region encompassing the gene is unfolded by translation of the upstream gene. (This structural arrangement probably evolved as a control mechanism assuring coordinated gene expression.) When ribosomes reach nonsense mutations near the 3′ end of the gene coding for the coat protein, the critical region of the RNA will have been unfolded, exposing the ribosome-binding site of the replicase gene and allowing its translation. These mutants will not, however, synthesize active coat protein, which is also the translational repressor of the replicase gene; hence replicase will be synthesized constitutively (more abundantly).

11.1. Both involve aberrant excision of an integrated DNA sequence, causing chromosomal DNA sequences to become associated with the emerging plasmid or phage genome.

11.2. (a) The information provided only suggests that β codes for the toxin. It is also possible that the phage codes for a factor that permits the synthesis of the toxin whose structure is specified by the bacterium.

(b) If the structural gene is indeed carried by the phage, one could demonstrate this with the help of a phage mutant that causes the production of an altered toxin protein upon lysogenization of non-toxinogenic bacteria.

11.3. The order is:

C ┤├ A ┤├ ———————————————— ┤├ B ├—

$\phi80$ phage

11.4. (a) The data suggest that the seven-kilobase sequence is a transposon which carried the gene for penicillinase. (b) The sequence (transposon) probably originated from a *Staphylococcus* plasmid, or else from a plasmid propagated in a third bacterial species. (c) Since the increase in penicillin is almost immediate following drug exposure, the enzyme appears to be substrate-inducible. (d) All phage particles in the rare plaque are resistance-conferring, because they arose from a single prophage which acquired the seven-kilobase sequence by transposition from a plasmid.

12.1. Newly synthesized DNA of the host has the template strand already modified, which is sufficient for protection.

12.2. (a) Phage X^- is a repressor mutant. Phage Y^- is an operator mutant.

(b) The rare plaques on strain C might reflect either a genotypic or a phenotypic phenomenon. (1) Host strain C might lack the proper receptor for phage ϕ. Therefore, the phage from the rare plaques might be host range mutants that have adapted to utilize a surface component of strain C as receptor. If true, these phage mutants would thereafter plate with high efficiency on strain C irrespective of their previous conditions of growth. Adsorption of the mutant phage to strain A, however, might be impaired. (2) Alternatively, host strain C has a DNA "restriction system" that destroys the unprotected phage DNA made in strain A. The rare plaques would be the result of accidental evasion of the entering phage genome from restriction; the DNA would then be replicated in a "modified" form which is thereafter protected. If this is true, phage particles from rare plaques would plate with high efficiency when directly retested on strain C, but would again plate at very low efficiency after the particles are regrown in strain A.

12.3.

Strain	Restriction	Modification
1	−	+
2	+	+
3	−	−

12.4.

DNA fragment used to infect the cells	β-galacto-sidase activity	Ampicil-linase activity	Growth on agar containing	
			Lactose	Ampicillin + glucose
Little Bam	+	−	−	−
Big Bam	−	+	−	+

12.5. Any number of obstacles might be met. For example: (1) Lack of a promoter that can be recognized by the E. coli RNA polymerase. (2) Lack of a positive regulatory protein for the expression of the toxin gene. (3) Clones expressing the toxin might be killed for unknown reasons. (4) Instead of being excreted, the toxin might be degraded by proteases either in the cytoplasm or in the periplasm.

12.6. In circular form, since a linear molecule would be cut into six fragments.

12.7. (a) The improved strains may have accumulated mutations in the penicillin biosynthetic system that increase the activity of key enzymes, decrease feedback inhibition or repression, increase enzyme synthesis (enhanced promoter efficiency of the genes, duplication of certain genes), and accelerate excretion of the antibiotic. (b) The strains might be further improved by collecting the biosynthetic genes into a multicopy plasmid so that the entire pathway can be amplified.

13.1. (a) Sensitivity may be tested either in liquid culture by the serial dilution method or on agar by placing various amounts of the drug on a filter disc. (b) Plasmid-coded detoxifying enzyme and reduced permeability of the bacterial cell membrane confer low level resistance. Rare mutations in the ribosome confer high level resistance. The basis of resistance can be tested by in vivo and in vitro drug inactivation, by cellular drug uptake, and by cell-free protein synthesis. (c) Prescribe other effective agents, such as isoniazid or kanamycin.

13.2. (a) It is useful to collect the following data first. S. aureus: pen^R (basis of resistance is excreted β-lactamase; sensitive cells are killed); spc^R (basis of resistance is intracellular; sensitive cells are growth arrested). *E. coli*: str^R (basis of resistance is intracellular; sensitive cells are killed); tet^R (basis of resistance is reduced permeability; sensitive cells are growth arrested).

(1) Streptomycin would kill S. *aureus*, but the action may be antagonized by tetracycline. *E. coli* would grow to 2×10^7.

(2) Growth of S. *aureus* would be arrested. *E. coli* would be killed until the β-lactamase (from S. *aureus*) inactivated the penicillin. Surviving *E. coli* would then grow.

(3) Growth of both organisms would be arrested.

(4) S. *aureus* would grow to 2×10^7. Growth of *E. coli* would be arrested.

(b) To test the number of each organism, one needs plates of selective media: a medium with spectinomycin for enumerating this S. *aureus*, and a medium with streptomycin for enumerating this *E. coli*.

13.3. Antibiotics might have lethal effects when they damage vital cellular components (such as ribosomes, DNA, and plasma membrane). For example, if all the ribosomes in a cell are irreversibly inactivated, there is no way for it to generate fresh and uninhibited ribosomes, even after the antibiotic is withdrawn. In contrast, the consequence would not be so severe if the irreversible inactivation involves all of the molecules of an enzyme, even an essential one (because the missing enzyme can be replaced by further synthesis). Thus reversible reactions of antibiotics will not be lethal, but irreversible reactions may or may not be lethal.

13.4. Statement 2.

13.5. (a) Increased cellular permeability as a result of penicillin action (even before killing) facilitates the entry of streptomycin which has intracellular ribosomal targets. (b) High level resistance is due to mutation in a specific ribosomal protein that alters (prevents) the binding of streptomycin to the ribosome. Low level resistance is due to an inactivating enzyme (plasmid-coded) or decreased permeability (chromosomal gene). (c) Certain mutational alterations in a ribosomal protein can cause conformational change that prevents protein synthesis. Binding of streptomycin further alters conformation in a compensatory way, allowing protein synthesis to proceed.

13.6. Streptomycin reduces the endogenous gut flora that normally compete for nutrients or produce inhibitory substances.

Index